T0230928

SUSTAINABLE DEVELOPMENT AND
GEOGRAPHICAL SPACE

Sustainable Development and Geographical Space

Issues of population, environment, globalization and education in marginal regions

Edited by

HEIKKI JUSSILA
University of Oulu, Finland

ROSER MAJORAL
University of Barcelona, Spain

BRADLEY CULLEN
University of New Mexico, USA

Routledge
Taylor & Francis Group

LONDON AND NEW YORK

First published 2002 by Ashgate Publishing

Reissued 2018 by Routledge
2 Park Square, Milton Park, Abingdon, Oxon OX14 4RN
711 Third Avenue, New York, NY 10017, USA

Routledge is an imprint of the Taylor & Francis Group, an informa business

Publisher's Note
The publisher has gone to great lengths to ensure the quality of this reprint but points out that some imperfections in the original copies may be apparent.

Disclaimer
The publisher has made every effort to trace copyright holders and welcomes correspondence from those they have been unable to contact.

A Library of Congress record exists under LC control number: 2001091440

ISBN 13: 978-1-138-74247-5 (hbk)
ISBN 13: 978-1-315-18231-5 (ebk)

Contents

List of Tables		*ix*
List of Figures		*xi*
List of Contributors		*xv*
Preface		*xvii*

1   Introduction   1
Heikki Jussila, Roser Majoral and Bradley Cullen

PART 1 – POPULATION AND SUSTAINABILITY

2   From unity to the present – spatial aspects of population   7
development in Tuscany seen on maps
Maria Andreoli, Heikki Jussila and Vittorio Tellarini

3   Patterns of population sustainability in the US Midlands   25
J. Clark Archer and Richard E. Lonsdale

4   Economic growth, ecological consequences and depopulation   37
in the rural areas of Galicia (Spain)
Rubén Loís-González

5   Local poverty in Finland in 1995   49
*Jarmo Rusanen, Toivo Muilu, Alfred Colpaert and
Arvo Naukkarinen*

6   Regional imbalances in Spain at the end of the century   65
Roser Majoral and Dolores Sánchez-Aguilera

7   Job development and regional structure – local examples of   83
growing and declining industries in Northern Finland
Toivo Muilu

8   Rural development and marginal areas in Slovenia   99
Stanko Pelc

PART 2 – ENVIRONMENT AND SUSTAINABILITY

9   Establishing a sustainable conservation policy in northern   115
    Saudi Arabia
    Gareth Jones and Ahmed Al Modayan

10 Biosphere reserves – sustainable development of marginal   129
    regions?
    Walter Leimgruber and Thomas Hammer

11 Sorghum based farming systems in Botswana – the challenges   145
    of improving rural livelihood in a drought prone environment
    Charles E. Bussing and David W. Norman

12 Global change and community self-reliance strategies in   159
    Southern Africa
    Etienne Nel

13 Subsistence and marginal lands – an Alaskan political and   173
    geographic issue
    Donald F. Lynch and Roger W. Pearson

14 Traditional water rights in northern New Mexico   186
    Olen Paul Matthews

PART 3 – GLOBAL ECONOMIC DEVELOPMENT AND SUSTAINABILITY

15 Spatial shifts in production and consumption – marginality   195
    patterns in the new international division of labour
    *Assefa Mehretu, Bruce WM. Pigozzi and
    Lawrence M. Sommers*

16 Economic change and transportation deregulation in selected   209
    marginal and critical areas of the US Pacific Northwest
    Steven Kale

17 Nearer the core – the fishing sector in one of Europe's marginal   223
    regions
    Xose Santos-Solla

18 Tourism and sustainable development – 'snow tourism' in   237
    European Mountain regions
    Francesco López-Palomeque and Marti Cors-Iglésias

19   Rural tourism and new patterns of development in the   249
     Portuguese rural space
     João Luís Fernandes and Fernanda Delgado Cravidão

20   Highly specialised production as an alternative in economically   261
     depressed areas – the wines of Priorat (Tarragona, Spain)
     Joan Tort-Donada

21   A tale right out of Hollywood – set in the desert of Almeira, in   270
     Spain?
     Hugo Capella-Miternique

22   Globalization and irregular urban growth from a Spanish   284
     example
     Maria José Piñeira Mantiñán

23   Sustainability, efficiency and economic success of RICA farm   295
     typologies in Italy
     Maria Andreoli and Vittorio Tellarini

PART 4 – EDUCATION AND SUSTAINABLE DEVELOPMENT

24   Higher education as a means to sustainable development in   315
     marginal regions
     Lennart Andersson and Thomas Blom

25   The development and implementation of sustainable education   335
     – northern areas, Pakistan
     Lucy Jones

PART 5 – CONCLUSIONS AND SUMMARY

26   Sustainability as seen from marginal regions – conclusions and   353
     summary
     Heikki Jussila, Roser Majoral and Bradley Cullen

List of Tables

Table 2.1   Number of communes in each population decile and   8
            total population of Tuscany, in 1861, 1931, 1951,
            1971, 1991, and 1998

Table 2.2   The largest municipalities of 1861 and 1998 and their   13
            populations, including those 'missing' in 1861 and
            1998

Table 2.3   The centres of growth and decrease in Tuscany, 1931   17
            to 1951 and 1951 to 1971, 5 largest based on absolute
            numbers

Table 2.4   Changes in population, Tuscany, 1981 to 1991 and   20
            1991 to 1998 (5 largest positive and negative
            communes)

Table 5.1   Decile limits, population by deciles, sample sizes and   51
            sample percentages

Table 5.2   Mean incomes subject to state taxation, average   53
            disposable incomes and income transfers (FIM) by
            decile in 1995

Table 5.3   Standard deviation of incomes by decile (FIM) in 1995   53

Table 5.4   Absolute distribution of grid cells and their population   59
            by deciles of average disposable income in 1995

Table 6.1   Basic economic indicators, Spain (1996-1997)   67

Table 7.1   Some basic facts on the municipalities of Sievi and   85
            Suomussalmi in 1997

Table 7.2   Some factors contributing to development differences   95
            between Sievi and Suomussalmi

Table 9.1   Details of protected areas, northern Saudi Arabia   119

Table 9.2   Problems faced by protected areas in Saudi Arabia   120

Table 9.3   Data layers used in the Saudi Arabian Environment   121
            Database

Table 10.1 The evolution of the biosphere reserve idea   132

Table 10.2 Population growth in the Entlebuch district, 1950-1990   136

Table 16.1 Population and employment changes in declining   212
            counties, 1970-1995

Table 18.1 Resorts, mechanical ski lifts and skiers in Europe, 1996   242

Table 22.1	Evolution of the GDP and employment structure, 1960-1992	290
Table 22.2	Development of the industrial activity around Vigo	290
Table 22.3	Economic structure of Vigo	292
Table 23.1	Main features of RICA sample in 1994: a) total values and b) average values	297
Table 23.2	Farm typologies from a private economic viewpoint	299
Table 23.3	Farm typologies according to economic performance and level of sustainability and efficiency	301
Table 23.4	RICA sample in 1994: farm typologies main features	302
Table 23.5	RICA sample in 1994: farm shares in terms of number, land and economic values according to typology	309
Table 23.6	RICA sample in 1994: farm distribution inside each typology according to geographical and altitudinal location	310
Table 24.1	Gross participation rate: higher education students, regardless of age, as a percentage of the population of the 5-year age group following the end of secondary schooling	316
Table 24.2	Measures to increase the number of new university students in the municipalities in Värmland	329
Table 25.1	Pakistan, five-year-plans 1995-1998: objectives and reality	338
Table 25.2	A structural analysis of major types of schools in the Northern Areas	340
Table 25.3	Project outputs and progress to date (June, 1999)	347

List of Figures

Figure 2.1   Concentration of population in Tuscany, 1861 and   10
1951, ranked according to the 1998 size of population

Figure 2.2   Concentration of population in Tuscany, 1971 and   11
1998, ranked according to the 1998 size of population

Figure 2.3   Population change in Tuscany from 1951 to 1971   12

Figure 2.4   The density of population in Tuscany, in 1861 and 1998   15

Figure 2.5   Population change in Tuscany from 1931 to 1951   16

Figure 2.6   Population change in Tuscany between 1971 and 1991   21

Figure 2.7   Population change in Tuscany from 1991 to 1998   22

Figure 3.1   Midlands study area   26

Figure 3.2   Landscape types   27

Figure 3.3   Population density, 1990   28

Figure 3.4   Population change 1960 to 1990   29

Figure 3.5   Interstate Highways   31

Figure 3.6   Metropolitan Status   31

Figure 4.1   Localization of Galicia in the Iberian Peninsula   38

Figure 4.2   Population density in Galicia, A) 1960, B) 1996   39

Figure 4.3   Distribution of bovine cattle head in Galicia, 1996   42

Figure 4.4   Distribution of porcine cattle head in Galicia, 1986   42

Figure 4.5   Distribution of European programs of rural   47
development in marginal areas of Galicia, 1999

Figure 5.1   Distribution of incomes subject to state taxation in   55
Southern Finland in 1995

Figure 5.2   Distribution of incomes subject to state taxation in the   56
Oulu region in 1995

Figure 5.3   Distribution of taxable and average disposable incomes   57
by distance from the centre of the local government
district concerned in 1995

Figure 6.1   Unemployment rate; Spain, 1991   68

Figure 6.2   Economic level; Spain, 1997   72

Figure 6.3   The location of the studied communities   73

Figure 7.1   Population development in the study areas in   86
1987-2000

Figure 7.2   Unemployment rates in the study areas in 1987-1997   88

Figure 7.3   Employed labour force resident in (a) Sievi and (b)   89
Suomussalmi by industries in 1987-1996

Figure 7.4   Employed labour force resident in study areas in 1987-   91
              1996: a) employed labour force; b) agriculture and
              forestry
Figure 7.5   Employed labour force resident in study areas in 1987-   93
              1996: a) manufacturing; b) public services
Figure 8.1   Population rank-size diagram for 48 Slovene cities   101
              according to the population size in 1869 and 1991
Figure 8.2   Potentially marginal rural areas consisting mostly of   108
              the settlements with unfavourable demographic
              conditions (from 1 to 10 according to the above text)
Figure 9.1   Main physical regions of Saudi Arabia   116
Figure 9.2   Areas of pasture in northern Saudi Arabia   118
Figure 9.3   Protected land, northern Saudi Arabia   119
Figure 9.4   Soil types of northern Saudi Arabia   122
Figure 9.5   Possible extension of preserved land, shown by dashed   124
              lines
Figure 9.6   Overlay of roads and settlements to assess conflict   125
              with conservation zones
Figure 9.7   Conflict map showing overlap between proposed   125
              conservation areas (dashed lines), good agricultural
              land (black areas) and good pasture land (grey area)
Figure 10.1   The landscape continuum   131
Figure 10.2   Model of the UNESCO biosphere reserve   133
Figure 10.3   The Entlebuch region in the Swiss context   135
Figure 10.4   Young and old age groups in %, 1950 and 1990   137
Figure 10.5   Location of potential biosphere reserves in Switzerland   142
              (1999)
Figure 13.1   Federal, state and native lands in Alaska   174
Figure 16.1   Counties declining in population and employment,   211
              1970-1995
Figure 18.1   Sustainable development of 'snow tourism': The role   245
              of governments
Figure 19.1   Provision of beds for rural tourism, 1997   257
Figure 21.1   Location of Almeria in Spain   271
Figure 21.2   Places related to the film industry in Almeria   278
Figure 22.1   Main urban areas in Vigo   287
Figure 22.2   Vigo and the expansion of the industrial activity   291
Figure 23.1   Northern, Central and Southern Regions of Italy   296

Figure 24.1   Level of education, percent of the population (aged 25-   318
                64) with at least upper secondary education and
                population with post-secondary education for some
                OECD countries
Figure 24.2   The proportion of new university students in the   331
                municipalities in Värmland, Sweden, in 1989/90 and
                1993/94, per thousand inhabitants
Figure 25.1   Location of Pakistan and its neighbours   336
Figure 25.2   Main physical regions of Pakistan   337
Figure 25.3   Administrative districts of Northern Areas   339

List of Contributors

Andreoli, Maria	University of Pisa, Pisa, Italy
Archer, J. Clark	University of Nebraska, Lincoln, NE, USA
Andersson, Lennart	University of Karlstad, Karlstad, Sweden
Blom, Thomas	University of Karlstad, Karlstad, Sweden
Bussing, Charles E.	University of New Mexico, Albuquerque, NM, USA
Capella-Miternique, Hugo	University of Barcelona, Barcelona, Spain
Colpaert, Alfred	University of Oulu, Oulu, Finland
Cors-Iglésias, Marti	University of the Balearic Islands, Palma de Mallorca, Spain
Cullen, Bradley	University of New Mexico, Albuquerque, NM, USA
Delgado-Cravidão, Fernanda	University of Coimbra, Coimbra, Portugal
Fernandes, João Luís	University of Coimbra, Coimbra, Portugal
Hammer, Thomas	University of Fribourg, Fribourg, Switzerland
Jones, Gareth	University of Strathclyde, Glasgow, Scotland, UK
Jones, Lucy	Educational Consultant, Glasgow, Scotland, UK
Jussila, Heikki	University of Oulu, Finland
Kale, Steven	Oregon Department of Transportation, Salem, OR, USA
Leimgruber, Walter	University of Fribourg, Fribourg, Switzerland
Loís-González, Rubén	University of Santiago, Santiago de Compostela, Spain
Lonsdale, Richard E.	University of Nebraska, Lincoln, NE, USA
López-Palomeque, Francesco	University of Barcelona, Barcelona, Spain
Lynch, Donald F.	University of Alaska-Fairbanks, Fairbanks, AK, USA
Majoral, Roser	University of Barcelona, Barcelona, Spain
Matthews, Olen Paul	University of New Mexico, Albuquerque, NM, USA
Mehretu, Assefa	Michigan State University, East Lansing, MI, USA

Modayan, Ahmed Al	King Abdulaziz University, Jeddah, Saudi Arabia
Muilu, Toivo	University of Oulu, Oulu, Finland
Naukkarinen, Arvo	University of Oulu, Oulu, Finland
Nel, Etienne	Rhodes University, South Africa
Norman, David W.	University of New Mexico, Albuquerque, NM., USA
Pearson, Roger W.	University of Alaska-Fairbanks, Fairbanks, AK, USA
Pelc, Stanko	University of Ljubljana, Ljubljana, Slovenia
Pigozzi, Bruce WM.	University of Michigan, East Lansing, MI, USA
Piñeira Mantiñán, Maria José	University of Santiago, Santiago de Compostela, Spain
Rusanen, Jarmo	University of Oulu, Oulu, Finland
Sánchez-Aguilera, Dolores	University of Barcelona, Barcelona, Spain
Santos-Solla, Xose	University of Santiago, Santiago de Compostela, Spain
Sommers, Lawrence M.	University of Michigan, East Lansing, MI, USA
Tellarini, Vittorio	University of Pisa, Pisa, Italy
Tort-Donada, Joan	University of Barcelona, Barcelona, Spain

Preface

This book is the outcome of research work done within the framework of the IGU Commission on Dynamics of Marginal and Critical Regions. The articles in this book stem from the annual conference of the Commission held in Albuquerque, New Mexico, USA in June 1999. The authors of the articles of this book have all worked in the fields of geography and regional economics and have done research that concentrate on the issue of marginality in the globalizing world. Following the four-year-programme of the Commission, the conference and, consequently, this book has the scope of analyzing the complex issue of sustainability from the economic, environmental, social and spatial points of view and hence the title: *Sustainable Development and Geographical Space*.

This title is indeed challenging and the editors are pleased to recognize that so many scholars were able to produce interesting and multi-level analyses of the phenomenon of sustainability in the field of spatial development. The editors of the book would like to express their gratitude to all those scholars that have contributed to this edition that is the fifth book produced by the Commission. The editors would also like to express their gratitude to Ashgate Publishing Ltd., that once again has given to the Commission the possibility to publish its research work in a book format.

The various aspects of sustainability discussed in this book show to the reader that the issues are indeed multitude and they range from purely economic aspects to the questions of education. The common theme and approach for all articles is that of a 'human' approach, by which we intend the softer more humanistic approach toward economic and social development issues. An approach that is common to those who do research in the environment of small local communities and/or marginal regions.

Finally, the editors would like to thank the organizers of the 1999 Albuquerque conference that brought together these scholars that made this book possible.

Pisa, Italy, Barcelona, Spain and Albuquerque, New Mexico, USA

Heikki Jussila, Roser Majoral and Bradley Cullen

1 Introduction

HEIKKI JUSSILA, ROSER MAJORAL AND BRADLEY CULLEN

The analysis of marginalized areas and peoples demonstrates that marginal regions world-wide are undergoing significant changes due to the volatility of political, social and economic processes both in developed (north) and less developed (south) areas. Parallel to this evolution, a general degradation of the ecosystem can be detected. Such changes are evident between macro and micro levels and at global, national, regional and minor civil division scales. Advancements in communication efficiency, entrepreneurial technology, and the shifting character of industrial organization are revolutionizing the process and structure of marginality in both north and south countries. Political unions (e.g. European Union), free trade and enterprise zones, transnational corporations and economic mergers of institutions (e.g. NAFTA) are impacting the nature of marginalization throughout the world.

While marginalization tends to increase with the current socioeconomic and political processes of globalization and deregulation, it is never a unidirectional process, as a look back into history since the Industrial Revolution will clearly demonstrate. It is also a very relative and complex concept that depends on the prevailing socioeconomic and political systems. The physical characteristics of an area also are subject to change due to peoples' perception and evaluation over time. This may denote a physically marginal region can become economically developed if its potential is recognized, and it can become depressed again when human preferences and/or other socioeconomic circumstances change (e.g. resource depletion). This means that sustainable development is becoming an important aspect for development in marginal regions. These regions, which traditionally have been those providing raw materials or labour, have been under overexploitation of resources and consequently today may suffer more from environmental, economic or social problems that prevent a sustainable use of local resources for the benefit of the region and its population. It is this that has prompted the need to analyze in 'Geographical Space' the aspects of sustainable development.

This book, entitled 'Sustainable Development and Geographical Space' is the fourth book in the series of books that are the product of research carried out under the research programme of the IGU Commission on 'Dynamics of Marginal and Critical Regions.' It is based on the research presented at the Commission's third annual meeting held in Albuquerque, New Mexico, USA (1999). The meeting in Albuquerque was the third annual meeting of the Commission. The two previous meetings were held in Harare, Zimbabwe (1997) and Coimbra, Portugal (1998). The topic discussed in the first meeting was 'Marginality seen on time' and the proceedings of this meeting were published as the second book of the present series, Marginality in Space – Past, Present and Future, edited by Heikki Jussila, Roser Majoral and Chris C. Mutambirwa. The theme for the Coimbra 1998 meeting was 'Globalization, Environment and Marginality'. The proceedings of this meeting consist of two volumes, 1) Globalization and Marginality in Geographical Space, 2) Environment and Marginality in Geographical Space, edited by Heikki Jussila, Roser Majoral and Fernanda Delgado-Cravidão. The last and concluding meeting of the Commission's four-year-research-period was held in Taegu, South Korea in conjunction with the 29th International Geographical Congress in Seoul, in August 2000.

The theme of the conference in Albuquerque was 'Sustainable development' and consequently all papers deal this question in one way or another. The question of sustainability is very complex and it can be interpreted in a variety of ways, as this volume clearly shows. However, the articles that in this volume are grouped under four main headings and conclusions, do bear some important similarities. These similarities arise from the fact that the contributions in this book all place the 'human action' in the foreground. This approach, typical for economic and human geographers, emphasizes the fact that to build a sustainable system the 'human' and indeed even 'humanistic' approach towards economic development is needed.

The first section, 'Population and sustainability', focuses on population, emphasizing the importance of a sustainable population base for economic development within a given region. The articles describe and analyze how population distributions, migration, employment, as well as the socioeconomic and demographic characteristics of various regions of the world impact the development of sustainable systems.

The second part of this book, 'Environment and sustainability', contains six chapters that analyze the conflict between the physical environment and society. The question of sustainability is linked to concerns about the environment. The articles in this section analyze the environmental ramifications of economic growth and present approaches to economic development that are environmentally sustainable.

The articles in the third section of this book, 'Global economic development and sustainability', discuss the influences of economic globalization on regional, national, and international effort to develop sustainable systems. The scale of the analyses extends from the global to the local. Efforts to establish systems that are globally or regionally sustainable can negatively impact local areas and *vice versa*. Increasingly, all scales of development are interconnected.

The last section of this book, 'Education and sustainable development', focuses on education's role in developing a sustainable system. The role of education in regional development is quite often forgotten, although it in many ways forms the basis for development. The lack of adequate system for education is a hindrance that can effectively 'stop' development. In this respect the two articles in this section, although different in their approach give an important contribution as one wants to comprehend the development possibilities of a specific region. Sustainability ultimately depends on the ability of society to utilize the earth's resources without destroying them. The prospects for success are linked to the level of education.

The editors hope that the following articles of this book give to the reader an overview of the issues that touch sustainability in marginal and critical regions. The research work presented in this book naturally is not able to include all possible aspects of sustainability. However, the variety of point of view that will be presented does show that this issues is been studied within the fields of economic, regional, rural and social geography intensively and from angles that do give an idea of how complex is the issue of sustainability in marginal regions. It is the hope of the editors that readers of this book will find the outlines and openings in this field imaginative and refreshing as well as informative.

PART 1
POPULATION AND SUSTAINABILITY

2 From unity to the present – spatial aspects of population development in Tuscany seen on maps

MARIA ANDREOLI, HEIKKI JUSSILA AND
VITTORIO TELLARINI

Introduction

Regional development of any kind is always dependent on multitude of 'variables' (e.g. ISTAT, 1989; Papi 1979; Andreoli, 1989). However, there are variables that are essential for sustainable economic and regional 'development' or 'underdevelopment' and the most important of these is quite naturally population (see Tellarini 1992, p. 198). Consequently most development programmes aim at 'enhancing' the conditions where people live or in the case of serious out-migration try to look for ways of 'stopping' or at least significantly reducing the level of out-migration from a specific region.

In this study a series of maps [1] is used for the analysis of population development of Tuscany. In addition to maps an analysis of the population concentration [2] patterns will be looked into in order to see if there are significant changes within the spatial distribution of population of Tuscany Region. The 1861-1998 population development of Tuscany has been analyzed from the viewpoint of regional sustainability and economic development of Tuscany region, while placing a special reference on the aspects of population movements within Tuscany.

A macro view: concentration of population, 1861-1998

Population development in Tuscany during the period 1861-1998 has gone through most of the changes Italy has experienced, which has meant:

1) The years of out-migration to the 'New World',
2) The problems of malaria in 'southern' coastal parts of Tuscany – the Province of Grosseto,
3) The time of having the Capital of Italy (Florence),
4) The First and the Second World Wars and subsequent 'baby booms',
5) The industrialization and
6) The development of a 'post-industrialized' society.

All these events have left their own mark on the population pattern of Tuscany.

Table 2.1 illustrates the concentration of population in Tuscany by measured in deciles and it shows that from after 1971 Tuscany has experienced a moderate decentralization; the number of municipalities making 50% of total population of Tuscany has increased. However, in macro terms population of Tuscany has been concentrating continuously, although Indeed, the ranking of population of Tuscany municipalities, based on the ranking of the year 1998 shows, the concentration pattern has not changed considerably, if one excludes extreme changes of 1861 curve (Figure 2.1).

Table 2.1   Number of communes in each population decile and total population of Tuscany, in 1861, 1931, 1951, 1971, 1991, and 1998

Decile	1861	1931	1951	1971	1991	1998
1	1	0	0	0	0	0
2	3	4	3	2	2	2
3	6	5	4	3	4	4
4	12	10	8	5	5	6
5	19	17	15	8	9	9
50%	41	36	30	18	20	21
6	25	23	22	15	15	16
7	31	30	29	23	22	22
8	41	40	40	35	34	34
9	55	56	58	56	52	51
10	94	102	108	140	144	143
Tuscany	1,920,577	2,913,965	3,158,811	3,473,097	3,581,051	3,528,563

Source: ISTAT and Tuscany Region http://www.regione.toscana.it/

Some municipalities have changed significantly in their ranking since 1861. However, this was to be expected as some municipalities on the coastal area of Tuscany were losing population during the late 1800s due to malaria, but the mountain and hilly areas of Tuscany gained in population. To give an example, the city of Grosseto had 4,724 inhabitants in 1861, less than the largest city of the Island of Elba, Portoferraio with 5,057 inhabitants. This explains for the large changes seen at the 'higher' end of the population pattern of 1861, when using the ranking of 1998.

During the first years of 1900s the concentration process continued (Figure 2.1). The growth of larger cities of Tuscany, Florence, Livorno, Prato and Pisa started during the first half of the 1900s. For instance, Florence, even taking into account the II World War, grew between 1931 and 1951 over 70,000 inhabitants. The previously sited 'exception' city, Grosseto, gained in population, although not as much as between 1861 and 1931 when it increased with almost 19,000 inhabitants and moved from the ranking 115 to 17 in 1931 and to 13 in 1951, with a total of 38,200 inhabitants. In spite of this process the situation for 1951 was – in terms of concentration – almost identical to that of 1931.

In Italy, like in most western European countries the 'great exodus' from the rural regions to cities[3] started after the World War II and in it mainly took place between 1951 and 1971, as can be seen from the maps describing population change from 1951 to 1971 (see Figure 2.3). At the time rural regions seemed to become 'empty' in a decade, but the congestion in cities has later partially reversed the flows, although the most 'marginal' areas have continued to loose population. The curves of concentration of population in 1971 and in 1998 (Figure 2.2) are almost 'identical' despite the time-difference of 27 years. There are, however, some differences in 1971 the last 10th decile comprised 140 municipalities and the 50% limit of population was reached by adding the 18 largest communes together. In 1998, the number of municipalities in the 50% group were 27, but the 10th decile consisted 143 municipalities. The concentration pattern seems to be slowing down slightly, while the number of municipalities with a very small population is increasing. These municipalities are located on the 'edges' of Tuscany and they can be deemed both 'rural' and 'marginal', increasing the spatial dimension of 'marginality'.

Table 2.1 showed how the decile pattern of municipalities has changed in Tuscany between 1861 and 1998. Table 2.2 shows the changes in the 10 largest municipalities of Tuscany. As in the case of Figures 2.1 and 2.2 the ranking of 1998 is used and consequently, there are municipalities that through changes in population have increased their 'importance' in the 'regional' and 'political' arena.

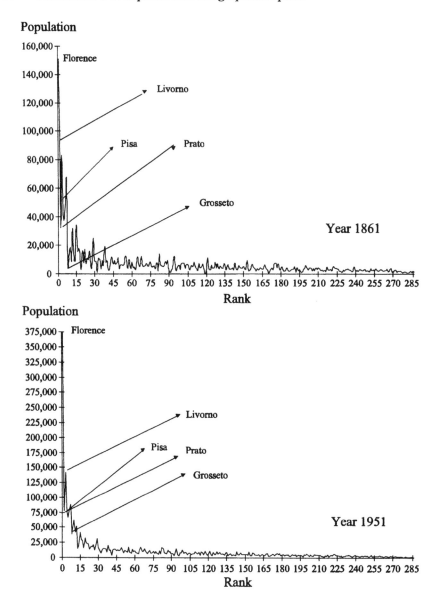

Figure 2.1   Concentration of population in Tuscany, 1861 and 1951, ranked according to the 1998 size of population

Source: ISTAT and Tuscany Region http://www.regione.toscana.it/

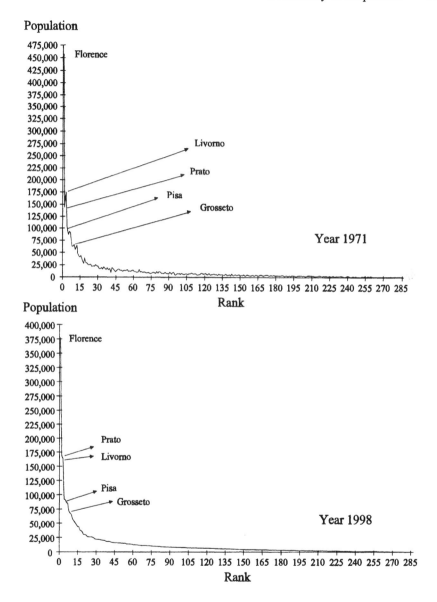

Figure 2.2   Concentration of population in Tuscany, 1971 and 1998, ranked according to the 1998 size of population

Source: ISTAT and Tuscany Region http://www.regione.toscana.it/

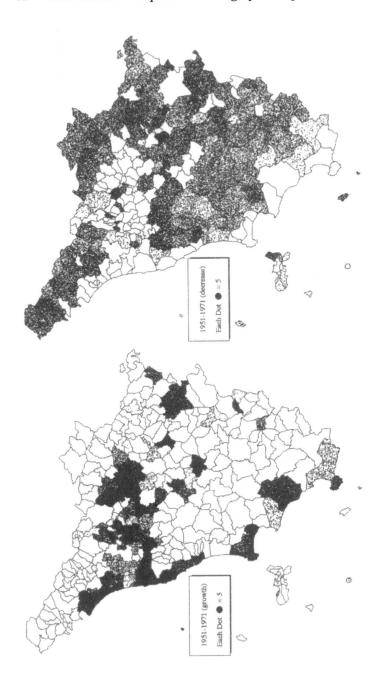

Figure 2.3   Population change in Tuscany from 1951 to 1971

A majority of the 10 largest municipalities in Tuscany have stayed the same throughout the whole period (Table 2.2). In 1861 the remaining municipalities were, in order of size Capannori 34,247, Siena 32,108 and Cortona 25,212 inhabitants. In 1998 the remaining municipalities were Grosseto 72,161, Massa 68,082 and Carrara 66,206 inhabitants. The ranking of these later mentioned municipalities in 1861 was for Grosseto 115, for Carrara 12 and Massa 17, while in 1998 Capannori had ranking 15, Siena 12 and Cortona 30.

Table 2.2   The largest municipalities of 1861 and 1998 and their populations, including those 'missing' in 1861 and 1998

Size ranking	Municipality	Year 1861	Municipality	Year 1998
1 (largest)	Florence	150,864	Florence	376,760
2	Livorno	94,977	Prato	162,321
3	Lucca	66,061	Livorno	171,135
4	Pistoia	49,584	Pisa	92,494
5	Pisa	43,835	Arezzo	91,301
6	Arezzo	37,100	Pistoia	85,906
7	Capannori	34,257	Lucca	85,558
8	Prato	32,710	Grosseto	72,539
9	Siena	32,108	Massa	65,692
10 (smallest)	Cortona	25,212	Carrara	68,005
Municipality	Grosseto (115)	4,724	Capannori (15)	43,739
missing in 1861	Massa (15)	15,319	Siena (12)	54,436
or in 1996	Carrara (12)	18,344	Cortona (29)	22,487
Population of 10		566,708		1,271,711
% of 10		29.51%		36.04%
Tuscany		1,920,577		3,528,563

Sources: ISTAT and Tuscany Region on Internet at http://www.regione.toscana.it/

Variations in time and space

Figures 2.1 and 2.2 show that population concentration in Tuscany has changed, especially between 1861 and 1951, but after the ranking order of municipalities has stayed remarkably constant, although the map information (Figures 2.3 to 2.7) shows clearly that the population has been concentrating on the more 'favourable' and 'accessible' regions. The changes in spatial distribution of the population in Tuscany are analyzed in four (4) time-periods:

1) From unity to beginning of 1930s,
2) From start of industrialization, and reconstruction after World War II,
3) Rural exodus and economic boom and
4) From emerging of service society to post-industrial and communications society.

From unity to beginning of 1930s

During the first period, from unity to early 1930s almost all communes, i.e., municipalities, of Tuscany gained in population and the population of Tuscany regions increased by almost a million (993,388) inhabitants and only 15 of the total of 287 communes (as of today) did loose population. Most of these communes were located in the province of Lucca (6 communes) and in the province of Livorno (5) in which the island communes did loose population. In the province of Lucca that contained most of the communes that did loose population already during this period where located in the area that currently forms the area of Communità Montana di Garfagnana. In that region, for instance the commune of Fosciandora has been loosing population throughout the whole investigation period.

During this period despite the almost 'universal' pattern of growth in population the largest changes occurred in the central valley of the River Arno and on the coastal Plain of Tirrenia. The absolute population distributions of the two extremes of the period, 1861 and 1998, show that the interior of the Tuscany in 1998 does not differ markedly in 'overall' density with respect to that of the 1861 (Figure 2.4).

From start of industrialization, and reconstruction after World War II

This second period of investigation (1931 to 1951) is not very good, since, the World War II, had a profound influence on population (Figure 2.5). This, however, is true throughout Europe and consequently the spatial distribution of population in Tuscany reflects the images obtained and seen in other European countries just before and immediately after the World War II. During this period the number of communes with negative population trend grew from 15 to 124. In spite of the increase in the number of 'negative signs' the population of Tuscany increased by almost quarter of a million (244,846) inhabitants in this period, which in spatial terms meant

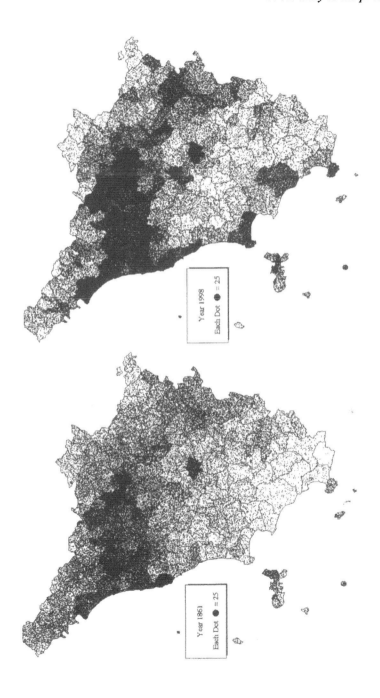

Figure 2.4   The density of population in Tuscany, in 1861 and 1998

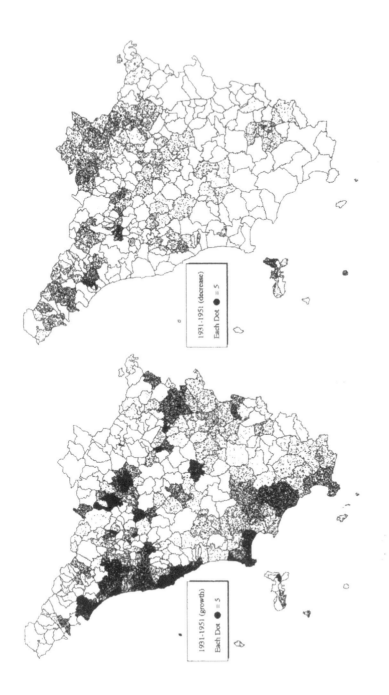

Figure 2.5   Population change in Tuscany from 1931 to 1951

an increase in population concentration, as the growth mostly occurred within or at the immediate proximity of the capitals, of the 10 provinces of Tuscany region. Table 2.3 provides additional information and 'proof' of the 'visual' analysis of the changes given in Figure 2.5.

Table 2.3 The centres of growth and decrease in Tuscany, 1931 to 1951 and 1951 to 1971, 5 largest based on absolute numbers

Commune	absolute	(%)	Commune	absolute	(%)
			Positive change		
Florence	70,465	18.8	Florence	83,178	22.2
Livorno	21,622	15.2	Prato	65,601	84.5
Prato	16,133	20.9	Livorno	32,458	22.8
Grosseto	15,102	39.5	Scandicci	32,326	213.9
Massa	10,719	21.3	Pisa	25,693	33.1
			Negative change		
Fivizzano	-2,617	-15.9	Cortona	-9,257	-29.0
Firenzuola	-2,332	-22,0	Castelnuovo Berardenga	-4,827	-48.6
Bagni di Lucca	-1,695	-14.7	Fivizzano	-4,774	-28.8
Barberino di Mugello	-1,686	-16.3	Firenzuola	-4,700	-44.3
Vicchio	-1,618	-15.5	Vicchio	-4,545	-43.4

According to the Table 2.3 Florence had in absolute numbers the highest growth between 1931-1951. However, as the percentage growth figures adjacent to the absolute numbers indicate, Florence did not have the highest relative growth. The highest relative occurred in the municipality of Capalbio in the province of Grosseto, with a 42.66% (not in the table 2.3). The capital of this province, Grosseto, also had a higher relative growth than Florence. In the negative end the commune of Rio nell'Elba had the highest relative decrease in population (-34.17%) – not in table, while the commune of Fivizzano experienced the highest absolute decline with -2,617 inhabitants (15.78%).

Rural exodus and economic boom

The real change and acceleration in the concentration to centres, the 'rural exodus', occurred in Tuscany during this period. The year 1951 marks the start of 'rural' exodus, common to many countries. This 'exodus' was in Italy, as in some other countries of Europe, e.g. in Finland, short and in some respects quite a 'violent' process that left many regions 'suffering' for a long period of time. Table 2.3 above tells a part of the 'story'. The story of tremendously strong growth in large centres and an equally high decline in the rural regions of Tuscany. The maximum growth rate of over 200% (213.87%) during a period of 20 years can be seen for Scandicci, a commune that is located immediately adjacent to Florence, and also another commune adjacent to Florence, Sesto Fiorentino with over 110% growth, showed that the 'drag' of the big city continued, although Florence itself had a growth rate only slightly above 22% (22.2%).

Figure 2.3 shows clearly how population has moved from the rural and mountainous areas to the Arno River Valley where the largest centres are located. This period of 'rural' exodus has had effects that can be seen even today as one aim to create new development. The 'wounds' of this exodus left in the minds of people are probably much deeper than most people think. It seems that somehow people feel that that this exodus wiped-out all resources and development forces from the rural regions and it is sometimes difficult to understand and remember that this 'exodus' has taken place 30 years ago.

From emerging of service society to post-industrial and communications society

During the 'great rush' to the centres the 'hard' values of industrialization were important. The arrival of service economy increased the importance of 'human' touch and consequently things and also values of people started to change. For rural regions and to the sustainability of regional development the service society provides a way to create news paths for development. Along with service society came large-scale tourism that included larger areas and started gradually include farms into the 'fabric' of tourism services, although the major effect of the service society in terms of jobs was in cities and major centres.

This type of development had an effect on the patterns of population and on the distribution of growth and decline areas. Figure 2.6 shows that people continued to move toward centres, but something had changed. The negative trends of Florence and Siena during the 1980s (the former -9,472 or -2.07% and the latter -3,645 or -5,55%) were the first signs of return migration. There was thus evidence that a high concentration of population had become, if not a liability and disadvantage, at least an 'undesirable' characteristic that should be 'avoided', if possible. In here, it is not possible to go deeper, but Gouérec (1995) while discussing the case of Garfagnana refers to this return phenomenon. Despite the 'feeling' of an unsatisfactory state of being, the process of concentration continued, e.g. the municipalities immediately adjacent to Florence – Scandicci and Sesto Fiorentino – gained in population. Scandicci grew by 6,597 (13.91%) inhabitants and Sesto Fiorentino grew by 3,461 (8.25%) between 1971 and 1981. In terms of growth, among the first 10, there were during this period six (6) communes of the province of Florence, all located relatively close to either Florence or Prato that had the highest growth rate in terms of absolute numbers 16,988 inhabitants. The highest relative growth rate of 61.19% belonged to the commune of Montemurlo that is located between Florence and Prato.

The growth of the Prato areas is a sign of the industrial and service development that expanded from Florence to this direction alongside with the main highway, autostrada, that leads from Florence to the cost, Livorno and Pisa and finally to Genoa. This communications link facilitated the location of industries, especially service and transport oriented, and thus the growth of the area has a logical and 'hard' economic reasoning.

The societal change from service society and later to a communications society has taken place gradually from 1980s onwards. The issue of population sustainability has become an issue for main cities that during 1980s and 1990s started loosing population in favour of their adjacent communes. In this respect the image is the same as during 1970s, although the distance from the main centre (Florence) to the growing communes did increase as can be seen from Figures 2.6 and 2.7. However, this phenomenon, so far has not occurred around other centres in Tuscany. Consequently, municipalities adjacent to the cities of Pisa, Livorno, Arezzo and Siena have gained in population, while the centre lost. The notable exceptions from the pattern of 'capoluoghi' that lost population are Prato and Grosseto (Table 2.4) who have had a positive population trend.

In the case of Prato the positive development is most probably due to the fact of industrialization that needs good communications routes, while in the case of Grosseto the growth is probably due to the expansion of tourism and consequently more in the line of 'post-industrial service' than in the line of communications. The evidence of this is that currently all adjacent municipalities including Castiglione della Pescaia, in the 1981 to 1991 period lost population, have changed pace and are now gaining in population.

Table 2.4   Changes in population, Tuscany, 1981 to 1991 and 1991 to 1998 (5 largest positive and negative communes)

Commune	1981 to 1991 absolute	(%)	Commune	1991 to 1998 absolute	(%)
		'Growing communes'			
Prato	5,487	3.42	Prato	5,428	3,28
Sesto Fiorentino	1,972	4.34	Campi Bisenzio	2,287	6,64
Grosseto	1,734	2.49	Cecina	1,611	6,54
Poggio a Caiano	1,655	26.33	Colle Val d'Elsa	1,572	9,23
Bagno a Ripoli	1,647	6.40	Montespertoli	1,532	16,24
		'Losing communes'			
Florence	-45,037	-10.05	Florence	-23,605	-5.85
Livorno	-8,229	-4.68	Pisa	-5,796	-5.86
Pisa	-5,581	-5.34	Livorno	-4,439	-2.65
Siena	-5,033	-8.12	Scandicci	-2,453	-4.58
Pistoia	-4,444	-4.82	Siena	-2,288	-4.02

When analyzing the population development the 'history' of Italian industrialization does come-up clearly. The most radical change has taken place during the period 1951 to 1971, when the Italian economy boomed and especially the northern Italy industrialized rapidly. Industrialization created new jobs and consequently there was a high degree of migration within Tuscany towards the centres of industry, like Pontedera, the home of 'Vespa Scooter', i.e., the main office of Piaggio.

During the 1980s and especially 1990s one can see signs of a new phenomenon connected to the 'saturation' effects of population concentration and urbanization. The 'return' migration of 1980s and 1990s can be observed, as 'rings' around larger centres. The population distribution and growth patterns reflect the changes that have taken place in the economic and social structure of the area, after all the way people think the future will look, influences their decisions whether to stay or to move.

Currently the two growing provincial capitals, Grosseto and Prato, possess different 'assets' for the growth pattern. Grosseto on the coast is developing tourism that has become increasingly important in the area. This in turn has created jobs and attracted tourists that look for a slightly 'different' environment than that of 'historically' most important tourist centre Florence. The Prato area, on the other hand, has developed based on a more traditional industrial pattern, but with good communications link,

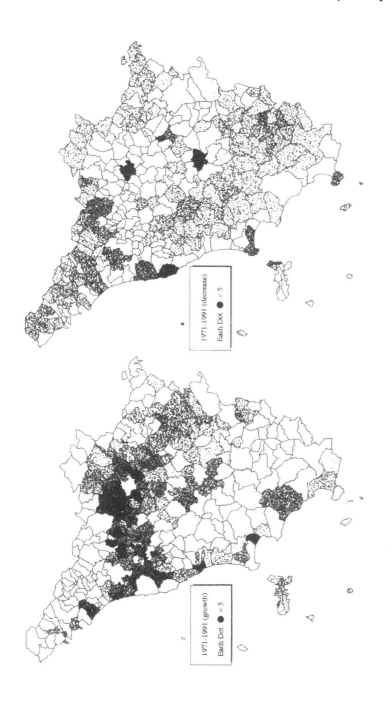

Figure 2.6 Population change in Tuscany between 1971 and 1991

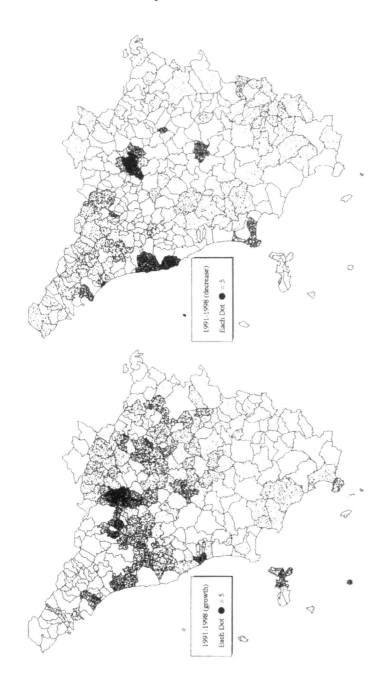

Figure 2.7   Population change in Tuscany from 1991 to 1998

the autostrada, that in Italy is more important for long distance transportation than in other European Union countries, where also railways play an
important part. The proximity to Florence has also helped in the sense that
connections to the Regional Capital are good.

Population in Tuscany has been moving toward centres, but as the
latest information shows at least Florence, with surrounding is now loosing
population. At the moment, it is only possible to hypothesize as has been
done earlier, that the affluent people have been moving. However, if this
hypothesis of affluent out-migrants holds, this development might in the
future create sustainability problems for large centres. An affluent countryside with good communications to the main centres may well be insight,
but as the negative signs of more remote communes show, this type of
development does not 'help' in solving the problems of these areas.

The population 'stability' of 1930s is currently gone and in order to get
both stability and sustainability it is necessary to go beyond the macro
images and look for the reasons for both metropolitan out-migration and
rural out-migration. These reasons are probably quite different in character,
but without this knowledge it is difficult, if not even impossible to develop
policies for sustainable spatial development. An aim that is common for at
least in 'theory' to nearly all policy efforts that seek enhancing local skills
or local economy. The regional policies of European Union, for instance,
will work only if there exists this 'knowledge' of the reasons of people, but
in order to make a difference the understanding of the macro phenomena,
like the one presented here is also needed. A reason of a person is the key
to understanding, but one should always try to see both the forest and the
tree before making a policy decision. In here the aim has been to look for
the forest, in order to find the trees that need to be examined more closely.

Combining the macro information and the local information, it is more
probable that one is also able to find a way to develop and start processes
that will lead to stability and through it to a 'more' sustainable development in a given region, since understanding population, however, is the key
to all development.

Notes

1. Due to the large number of maps available (more than 20) only some (5) are presented
    in this article.
2. The concentration of population is analyzed through time-based rank-size diagram, that
    collects the most relevant years and tables of concentration of the 5 largest and smallest
    municipalities of Tuscany.

3. In Tuscany this meant the cities of Florence, Livorno, Pistoia, Prato and Pisa and later also Grosseto, Arezzo and to some extent also Siena.

References

Andreoli, M (1989), 'Toscana, in Cannata', G. (ed.), *I sistemi agricoli territoriali italiani,* Franco Angeli, pp. 288-311.

Gouérec, N. (1995), *La Communità montana della Garfagnana, ovvero il facino discreto della marginalità?* Communità Montana della Garfagnana and Università di Pisa, Dipartimento Economia dell'Agricoltura, Pisa (mimeo).

ISTAT (1986), 'Classificazione dei comuni secondo le caratteristiche urbane e rurali', *Note e Relazioni* 2, 1986.

ISTAT (1996), *Popolazione residente dei comuni. Censimenti dal 1861 al 1991, circoscrizioni territoriali al 20 Ottobre 1991,* ISTAT, Roma.

Papi, Giorgi (1979), 'Un tentativo di classificazione dei comuni toscani mediante il metodo delle componenti principali', *Note Economiche,* 2-3.

Tellarini, V. (1992), 'Some questions about socio-economic marginality in rural areas of developed countries', in Gade, Ole (ed.), *Spatial Dynamics of Highland and High Latitude Environments,* Vol. 4, 1996, Occasional Papers in Geography and Planning, Appalachian State University, Boone, pp. 193-210.

Tuscany Region, http://www.regione.toscana.it/, The Official Internet Site of the Regione Toscana, Population Statistics.

3 Patterns of population sustainability in the US Midlands

J. CLARK ARCHER AND RICHARD E. LONSDALE

Introduction

The population maps of seemingly every country change constantly as people rearrange themselves across the landscape. Some regions are able to expand or at least sustain their population sizes, while others experience net loss. As for marginal lands, the role of population change in characterizing and delimiting them has been well established (Sanchez-Aguilera, 1996; Lonsdale and Archer, 1999). In general, marginal lands are apt to suffer net out-migration, population loss, and an ageing of the remaining population.

With sustainability a major theme in this book, it is appropriate to link this concept to the matter of population change. If sustainability is viewed 'from the perspective of the development of a long enduring societal system', as Miller (1998) has suggested, it can be proposed that a sustainable population level is one which will permit a societal system to be long enduring. The purpose of this study is to examine the extent to which certain areas are able or unable to sustain their populations and why.

The study area

A 15 state region, here labelled the U.S. Midlands (Figure 3.1), provides a broad laboratory in which to consider several independent geographical variables likely to influence population trends. The Midlands cover 40 percent of the area of the U.S. and embrace 54.9 million people, one-fifth the U.S. total. Research maps were prepared using county level population data (U.S. Bureau of the Census, 1993; 1996; 1998) matched to locational references using Atlas*GIS (Strategic Mapping, 1992).

Figure 3.1   Midlands study area

The natural landscape of the Midlands (Figure 3.2) is dominated by plains interrupted in the west by the Rocky Mountain System (Rockies) and in the east by the Ozark-Ouachita Uplands (Ozarks). Average annual precipitation exceeds 100 cm. in the south-east, but drops under 50 cm. in the west. Winters are severe in the north, but states in the southern part of the study area are regarded as part of the 'sunbelt' with mild winters. Not surprisingly, population densities vary widely (Figure 3.3).

The Midlands as a whole experienced substantial population growth between the 1960 and 1990 census years, expanding 39.6% from 36.1 to 50.4 million, closely paralleling the national U.S. pace. The upward trend continued in the 1990s, to reach an estimated 54.9 million in 1997. However, there were great geographic disparities in this growth (Figure 3.4). Of the Midlands' 1264 counties, the 173 designated as metropolitan (metro) cornered 89.5% of the gain. The 426 non-metro counties adjacent to metro ones (non-metro adjacent) claimed 8.5% of the population increase, and the remaining 661 non-metro nonadjacent counties shared just two percent of the overall gain. Indeed, 602 counties of dominantly non-metro nonadjacent designation lost population between 1960 and 1990.

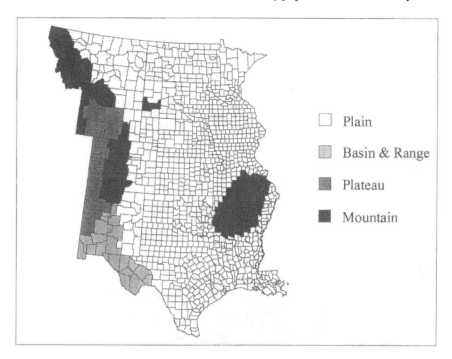

Figure 3.2   Landscape types

Some factors promoting population sustainability

Speedier transport and increased leisure time seem to have enhanced the attraction of the majestic scenery and abundant recreational opportunities offered by mountains. The Rockies and to some extent the Ozarks have long been popular tourist destinations. But until about 1960 few outsiders considered permanent or seasonal residence there because of the limited infrastructure and an employment base confined largely to marginal farming or ranching, logging, mining, and seasonal tourism. After 1960, however, obstacles to mountain residency diminished as Americans became more affluent and accessibility by automobile and aeroplane improved dramatically. In the 30 years after 1960 the population of the Midlands' 214 upland counties (Rockies and Ozarks combined) increased by 70.3%, from 4.4 to 7.5 million. The gain was most impressive in upland metro counties (98.7%), but also good in upland non-metro adjacent (52.3%) and even in upland non-metro nonadjacent (45.7%) counties.

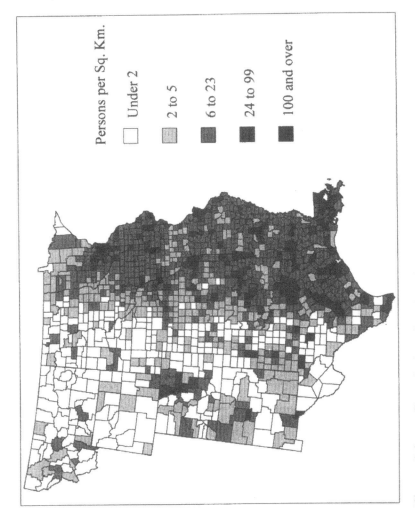

Figure 3.3   Population density, 1990

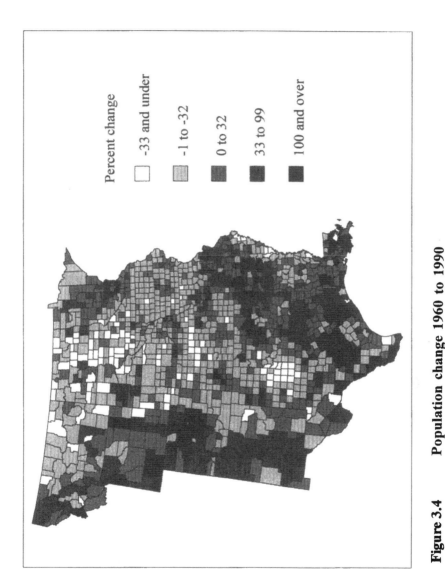

Figure 3.4 **Population change 1960 to 1990**

Much of the expansion in the Rockies was concentrated in Colorado and New Mexico, with smaller gains in Montana, Wyoming, and South Dakota (whose Black Hills are considered an outlying part of the Rockies). Of special note is the suburban and ex-urban expansion into mountain hinterlands from the rapidly growing urban corridor centred on Denver at the foot of the Rockies (Riebsame, 1997). A similar corridor is found in New Mexico, centred on Albuquerque. Sharing in the growth frenzy have been a great many non-metro counties rich in physical amenities and having developable (non-government) land available for building homes for seasonal or year round residence. The Rockies abound in tourist resorts, artist colonies, corporate retreats, and the like. As Rudzitis (1993) put it, 'the physical landscape provides the region's real economic base'.

In the case of the Ozarks, population growth constituted a startling reversal of earlier trends. For many decades prior to 1960 the Ozarks had been a classic marginal region, experiencing poverty, net out-migration, and population loss. The region was transformed in part because of Federal reclamation projects, which created several large lakes in Missouri and Arkansas and spurred a major expansion of recreation and tourism. Many planned retirement/recreation communities began to be developed, attracting people of varying financial means primarily from urban centres in the North (Siegel et al., 1995). Many of these people have been 'returnees' coming home to retire (Campbell, 1993). Bender *et al.* (1985) classified over half of the 110 Ozark counties as 'retirement counties'.

When plans were announced in 1956 to construct an Interstate Highway system interconnecting US metropolitan centres, residents of many non-metro counties hoped one of the highways would pass their way and bring employment opportunities and population growth (Figure 3.5). As it turned out when the system was largely completed in the early 1970s, 746 Midlands counties were positioned on or near an Interstate highway (county centre within 24 km), and 518 counties were more distant. Among the non-metro nonadjacent counties, those on or near an Interstate highway saw a 7.9% population increase between 1960 and 1990, while those more distant saw only a 1.3% gain (Figure 3.6). Thus, proximity to the Interstate Highway system has affected population sustainability in the Midlands.

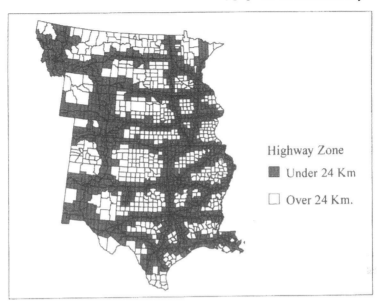

Figure 3.5   Interstate Highways

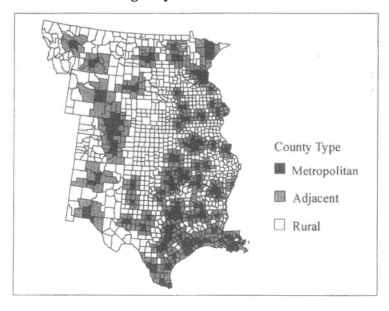

Figure 3.6   Metropolitan Status

Population density in itself can be a factor promoting population sustainability. The pattern shown in Figure 3.3 is largely a function of differences in average annual precipitation, i.e. heavier in the east, the location of metropolitan counties and more populated non-metro counties along Interstate highways, and proximity to the Gulf of Mexico. These more densely populated areas have more diversified economies and thus are more likely to sustain their populations (Case and Alward, 1997). Within the plains portion of the Midlands, the lowest population densities in 1990 were in the non-metro nonadjacent counties (6.1 persons/sq. km.), and a great many could not sustain their sparse populations. The non-metro adjacent counties (9.7 persons/sq. km.) experienced modest population increases, while plains metro counties (138.6 persons/sq. km.) enjoyed substantial growth. In the Rockies and Ozarks the population density population change relationship breaks down, as both low density and high-density counties shared in the post 1960 population surge.

The presence of government facilities is often helpful in sustaining population. Counties with state capitals show up well in the population change map (Figure 3.4). The same can be said for counties with state colleges and universities. Penal institutions are increasing in number and are eagerly sought by counties with lagging populations. At the Federal level, military bases have played a major role in many Midlands counties, though recent base closings or downsizings have made them a less reliable employer. In New Mexico the development of nuclear weapons and missiles has been a critical factor in the employment base of several counties, and New Mexico leads the nation in per capita Federal spending.

The post World War II expansion of irrigated agriculture has positively impacted population patterns, particularly in the drier plains where the Ogallala aquifer extends from west Texas to northern Nebraska. Large scale irrigation commenced along the Texas border with New Mexico in the 1950s, supporting cotton and feed grain production (Roberts, 1996). Western Kansas followed suit in the early 1960s, focusing on feed grains (White, 1994), and large areas of Nebraska soon followed a similar course. These developments were made possible by improved pumping technologies and the availability of inexpensive electric power from rural electric co-operatives (Rhodes and Wheeler, 1996). The positive impact on population was first seen in the labour intensive cotton areas of Texas, but later the irrigated areas producing feed grains attracted cattle feedlot operations and then large meat packing plants. The latter had to recruit large numbers of immigrant workers. The Gulf coast of eastern Texas and Louisiana also experienced increases in irrigation (Riebsame, 1997), primarily for rice production.

Mining development has been significant in a number of Midlands counties since 1960. Of note have been operations involving low sulphur coal in west central Colorado, north-west Wyoming, and adjacent areas in Montana; and the extraction of oil and gas along the Louisiana coast.

Areas with high concentrations of minority groups often experience expanding populations, even despite high unemployment and poverty (Lonsdale and Archer, 1999). The heavily Hispanic areas of the Lower Rio Grande Valley, southern and north-eastern New Mexico, and southern Colorado show high birth rates and high natural population increase, together with in-migration from Mexico (Saenz and Ballejos, 1993). Native Americans on or near Indian Reservations in New Mexico and northern Plains states also have high birth rates and high natural population increase, plus a marked reluctance to out-migrate (Lewis, 1998). These areas are apparent on the map of population change (Figure 3.4). However, the heavily black areas in the Lower Mississippi Valley of Arkansas and Louisiana do not show population growth because of considerable net out-migration.

Factors working against population sustainability

Insufficient employment diversity characterizes the great majority of Midlands counties losing population. Most are heavily dependent on agriculture with a minimum of attractive alternative employment opportunities. U.S. output per farm worker increased 13 fold between 1940 and 1989 (Rathge and Highman, 1998). The Midlands' farm population correspondingly decreased from 11.1 million in 1930 to 4.7 million in 1960 and 1.5 million in 1990. The rate of farm population decline from 1960 to 1990 (67%) has been fairly uniform all across the Midlands. Farm population densities per sq. km. are now remarkably low (e.g. in 1990 only 0.7 in the plains and 0.4 in the mountains and plateaux; and 0.8 in areas receiving more than 50 cm. average annual precipitation and 0.2 elsewhere).

Some of the Midlands counties losing population are ones overly dependent on mining. While coal, oil, and gas operations have stimulated growth in some areas, other mining activities have diminished. For example, in north-eastern Minnesota new processing procedures reduced labour needs in iron ore/taconite mining and processing, and in western Montana major copper operations virtually ceased in 1980 (Borchert, 1987). Oil shale deposits in sparsely populated western Colorado and south-western Wyoming sparked feverish speculation after the 1980 Federal Synthetic Fuels bill (Dietz, 1984). Local employment and population soared. Then Federal support waned in 1982 and oil shale development was all but abandoned. Unexpectedly, housing built for workers in upland settings proved attractive to retirees and second home purchasers, so that oil shale counties show strong 1960 to 1990 population gains in Figure 3.4.

Many of the counties overly dependent on agriculture or mining have tried unsuccessfully to diversify their economies. An estimated 46% of U.S. communities under 2500 in population have formed development committees to seek new industry or other businesses (Johansen and Fuguitt, 1984). In the 1960s and 1970s many Midlands rural communities, mostly in the eastern half of the region, did attract usually small manufacturing plants seeking modest wage non-union labour, and this served to retard or reverse population loss for a time. However, many of the workers displaced from agriculture or mining chose to migrate to a metro centre or out of the region altogether. Young people continued to do the same on graduation from high school. After about 1980 the trend to globalization of production made it even more difficult for most of these communities to expand their employment base.

Another critical factor in the inability of many counties to sustain their population is the general perception that they lack both physical and cultural amenities. In effect, they have few if any of the features which have attracted people to growing sections of the Midlands: mountains, ski slopes, lakes, the seashore, mild winters, superior shopping facilities, universities, professional sports, etc. In addition, business executives who make key decisions on where to locate new plants or research labs are also influenced by amenities as well as by such essential matters as labour supply, transportation facilities, access to markets, and availability of support services that metropolitan centres can provide. Non-metro counties losing population have their own rural style amenities, but this has been insufficient to turn the tide of population decline.

Summary thoughts

The ability of an area to sustain or expand its population is directly related to the diversity of its employment base, and this in turn usually requires proximity to a larger urban centre. Since amenities are so important to Americans in their locational choices, the most highly favoured places to live and work are the metropolitan areas (more specifically their suburban and ex-urban environs) situated amidst or adjacent to mountains, lakes, seashores, and the like. Those seeking retirement or seasonal recreation have more locational flexibility, but still like to be reasonably close to a metro centre. Thus, areas lacking both amenities and employment diversity are apt to experience ongoing demographic retrogression.

The notion that population sustainability or growth permits a societal system to endure might well be questioned. To be sure, the future looks bleak for sustaining the family farm societal system on the plains where the population is declining. But does this imply that societal systems elsewhere (e.g. those in the pre-1960 Rockies) with rapidly expanding populations will endure culturally? These places are being impacted by hundreds of thousands of new residents, and it may well be that new societal systems are being created or superimposed on existing ones.

References

Bender, Lloyd D. *et al.* (1985), *The Diverse Social and Economic Structure of Nonmetropolitan America*, Rural Development Research Report 49, U.S. Department of Agriculture, Economic Research Service, Washington, D.C., pp. 17-19.

Borchert, John R. (1987), *America's Northern Heartland*, University of Minnesota Press, Minneapolis, Minnesota, pp. 178-182.

Campbell, Rex R., Spencer, J.C. and Amonker, R.G. (1993), 'The Reported and Unreported Missouri Ozarks', in Lyson, Thomas A. and Falk, William W. (eds.), *Forgotten Places: Uneven Development in Rural America*, University Press of Kansas, Lawrence, Kansas, pp. 30-52.

Case, Pamela and Alward, Gregory (1997), *Patterns of Demographic, Economic and Value Change in the Western United States*, Report to the Western Water Policy Review Advisory Commission, U.S. Department of Agriculture Forest Service, Washington, D.C., p. 16.

Dietz, John (1984), 'The Colorado Shale Rush of 1981', *Great Plains-Rocky Mountain Geographical Journal*, vol. 12, pp. 47-59.

Johansen, Harley E. and Fuguitt, Glenn V. (1984), *The Changing Rural Village in America*, Ballinger, Cambridge, Massachusetts, p. 165.

Lewis, David R. (1998), 'Native Americans: The Original Rural Westerners', in Hurt, R. Douglas (ed.), *The Rural West Since World War II*, University Press of Kansas, Lawrence, Kansas, pp. 12-37.

Lonsdale, Richard E. and Archer, J. Clark (1999), 'Demographic Factors in Characterizing and Delimiting Marginal Lands', in H. Jussila, R. Majoral, and C. Mutambirwa (eds.), *Marginality in Space: Past, Present and Future*, Ashgate, Aldershot, pp. 129-143.

Miller, Vincent P. Jr. (1998), 'Sustainability, Regionalism, and the Trade-Offs of Development', in Lennart Andersson and T. Blom, (eds.), *Sustainability and Development, Regional Science Research Unit*, University of Karlstad, Karlstad, pp. 258.

Nord, Mark (1997), 'Overcoming Persistent Poverty B and Sinking into It', *Rural Development Perspectives*, vol. 12, no. 3, pp. 2-10.

Rathge, Richard and Highman, P. (1998), 'Population Change in the Great Plains', *Rural Development Perspectives*, vol. 13, no. 1, pp. 19-26.

Rhodes, Steven L. and Wheeler, S.E. (1996), 'Rural Electrification and Irrigation in the U.S. High Plains', *Journal of Rural Studies*, vol. 12, pp. 311-317.

Riebsame, William E. (1997), *Western Land Use Trends and Policy, Report to the Western Water Policy Review Advisory Commission*, Department of Geography, University of Colorado, Boulder, Colorado, pp. 32-56.

Roberts, Rebecca (1996), 'Recasting the Agrarian Question: The Reproduction of Family Farming in the Southern High Plains', *Economic Geography*, vol. 72, no. 4, pp. 398-404.

Rudzitis, Gunnars (1993), 'Nonmetropolitan Geography, Migration, Sense of Place, and the American West', *Urban Geography*, vol. 14, no. 6, pp. 576.

Saenz, Rogelio and Ballejos, M. (1993),'Industrial Development and Persistent Poverty in the Lower Rio Grande Valley', in Thomas A. Lyson, and William W. Falk (eds.), *Forgotten Places: Uneven Development in Rural America*, University Press of Kansas, Lawrence, Kansas, pp. 102-124.

Sanchez-Aguilera, Dolores (1996), 'Evaluating Marginality through Demographic Indicators', in Singh, R.B. and Majoral, R. (eds.), *Development Issues in Marginal Regions*, Oxford and IBH, New Delhi, pp. 133-148.

Siegel, Paul B., Leuthold, F.O., and Stallmann, J.I. (1995), 'Planned Retirement/Recreation Communities Are Among Development Strategies Open to Amenity-Rich Areas', *Rural Development Perspectives*, vol. 10, no. 2, pp. 8-14.

Strategic Mapping (1992), *Atlas*GIS*, Strategic Mapping, Santa Clara, California.

United States Bureau of the Census (1993), *1990 Census of Population and Housing*; Summary Tape File 3C: United States Summary on CD-ROM, United States Bureau of the Census, Washington, D.C.

United States Bureau of the Census (1996), *USA Counties 1996 on CD-ROM*, United States Bureau of the Census, Washington, D.C.

United States Bureau of the Census (1998), *County Population Estimates for July 1, 1997* (via Internet from WWW.census.gov), United States Bureau of the Census, Washington, D.C.

White, Stephen E. (1994), 'Ogallala Oases: Water Use, Population Redistribution, and Policy Implications in the High Plains of Western Kansas, 1980-1990', *Annals of the Association of American Geographers*, vol. 84, no. 1, pp. 29-45.

4   Economic growth, ecological consequences, and depopulation in the rural areas of Galicia (Spain)

RUBÉN LOÍS-GONZÁLEZ

Introduction

A brief look at the map (Figure 4.1) reveals that Galicia should be considered an accident. Due to a variety of historical events and accidents, this territory in the Northwest of the Iberian Peninsula was incorporated into Spain and not into Portugal. This incorporation – which gradually consolidated during the 17th, 18th and 19th centuries – turned Galicia into a peripheral, isolated, backward region, remarkably rural, compared to the State it now officially belongs to (Beiras, 1972; García Fernández, 1975). Furthermore, Galicia, together with a few other northern Spanish territories such as Asturias, Cantabria and the País Vasco, constitutes the Atlantic world in an over-whelmingly Mediterranean context. Its lush vegetation and plentiful water resources distinguish Galicia from a larger area where these are scanty. Besides, the supremacy of small agricultural property sets Galicia apart from an area in Southern Europe where large property predominates, or where at least there are some contrasts (Méndez and Molinero, 1993).

All these features account for the fact that Galicia remained, until very recently, the main rural area in Spain. In 1960, more than two thirds of the Galician active population still belonged to the primary sector, and only 27% of inhabitants dwelled in urban centres, rural densities (Figure 4.2) generally ranging from 50 to 80 inhabitants/km^2 (Bertrand, 1992; Lois and Santos, 1995; López Taboada, 1996). However, in spite of constituting the most numerous group of Galician population, agricultural workers were not productive from

Figure 4.1   Localization of Galicia in the Iberian Peninsula

a statistical point of view, since all the goods from agriculture, cattle breeding, forestry and fishing were for subsistence. This system, based on small family-owned properties, ranging from 5 to 6 hectares, has undergone drastic changes within the last forty years. The scope of these changes, which have altered Galician society and space, will be taken up later in this essay. Nevertheless, I would like to advance two general facts, both revealing and contradictory. On the one hand, the final Agrarian Production in Galicia has become more than twenty times larger than in 1960 (81.1 million ECU in 1960 and 1680.1 in 1994). On the other, this improvement has taken place in 15-20% of the land available, in keeping with the abandonment of the land or its mere transformation into forest (Lois and Santos, 1995; López Iglesias, 1996; Lois and Martínez 1998).

Figure 4.2   Population density in Galicia, A) 1960, B) 1996

The crisis of traditional agricultural society and the consolidation of competitive agriculture in a global context

In the 1950s and 1960s, when the Galician population clustered in small villages (about 30,000) scattered around the whole area, its main goal was to produce enough goods to cover basic food and clothing needs. A commercial type of agriculture did not exist, because the Spanish urban market was distant and weak, in the aftermath of a Civil War (1936-39) and a in post-war era characterized by international isolation. Families managed to survive by strategically keeping their best land for cereals, in order to make bread (mainly out of rye and corn), and for potato crops. Meat was obtained from the domestic breeding of a few pigs. Flax and sheep wool allowed peasants not to depend on the outside for textile raw materials. The meadows and the mountain played a secondary role in the organization of agrarian space. This was due to the fact that stockbreeding and livestock by-products (cows, calves, mules and cheese),which were clearly market-oriented, were only considered as complementary to the main activity, as simply a means to obtain certain amounts of money from their sale in fairs or nearby markets. The whole system relied on the intensive growth of multiple crops and the work of a large number of hands.

Outside dependence was limited to the purchase of salt (essential for the preservation of goods), since honey replaced sugar, animal fat replaced vegetable oil, and the rest of the products were expendable. The economy opened itself up to the outside in a few cases (some areas sold wine or fish), but it was invariably independent (Torres, Lois and Pérez Alberti, 1993).

The irreversible crisis of the traditional peasant way of life took place in a short period of time, between 1960 and the early 1980s. The industrialization process and the intense economic growth of Spain, along with the gradual opening of Galician rural areas to the outside, and migration opportunities for the youth, account for such a quick transformation. In fact, a demographic decline occurred in the Galician countryside from 1951 to 1975. The regional migration rate was minus 455,680 people, 17.5% of the total population. In interior rural areas this percentage was as high as 25-30% (Hernández Borge, 1992). These generations, with age groups ranging from 15 to 30, who had not been able to migrate due to the economic crisis affecting Spain and Europe in the 1940s and early 1950s, gave way to younger generations who massively left their native villages in search of better job opportunities in industrial and urban labour-seeking regions or countries (Basque Country, Catalonia, Madrid, Germany, Switzerland, France, etc). Such a large-scale, accelerated rural exodus provoked a shift from relative rural overpopulation to a shortage of agrarian labour and, within a few years, it triggered a severe demographic decline in extra-urban areas. This demographic drain was in turn paralleled by the city's increasing demands of inexpensive meat and milk and, to a lesser extent, vegetables, wine, etc. Thus, in the 1960s and 1970s, the Galician agrarian broke away from multiple crops and subsistence farming, and launched specialization. Specialized production aimed at an output that was competitive both in the Spanish and European markets. It generally focused on milk and meat bovine, but also pork, poultry, tomato, vineyard, potato, etc (Pérez Touriño and Colino, 1983).

Presently, Galician rural areas continue to lose inhabitants, but some kind of commercial agrarian activity has been established in those areas in accordance to their possibilities. Thus, the rural areas of Lugo, Pontevedra and A Coruña specialize in bovine cattle, due to better conditions and good bioclimatic possibilities for meadows and fodder. The almost one million bovine cattle heads bred in these lands are used for to the production of milk. However, farmers also focus on the improvement of an endemic breed 'rubia galega', known for its good quality meat. Flax and cereal crops have ceased production, being replaced by grass lands and other cattle-oriented crops. Even though commercial cattle breeding has impoverished the landscape, it has promoted a considerable economic growth. Other areas with a more or less focal distribution, such as some parts of Ourense and Sarria have chosen to specialize in pork and poultry farming. Different areas focus on different activities. For instance, vineyards stand out in the regions with a controlled appellation (Ribeiro, Valdeorras, Rías Baixas, etc); potato are grown in specific districts like A Limia; pine and eucalyptus forestry proliferate close to the coast (Noia, A Barcala, Ortegal, etc); and horticulture is characteristic of the areas that are relatively close to the main cities. This new situation is the result of a shift from a multiple-crop subsistence system, to a specialized system (cattle breeding, commercial agriculture, forestry). As a consequence of this shift, the landscape has been simplified. The transition from a peasant to a purely commercial agriculture has been effected in a time span smaller than a generation, without rural property having significantly changed hands (most of the plots of land cover less than 10 hectares). However, the labour shortage has caused production to concentrate on a few areas (Lois and Santos, 1995).

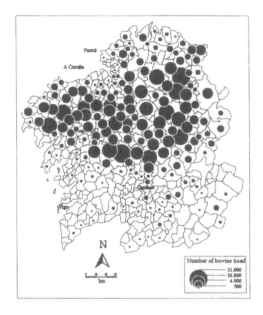

Figure 4.3   Distribution of bovine cattle head in Galicia, 1996

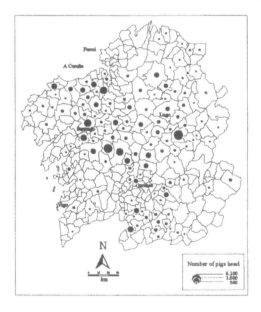

Figure 4.4   Distribution of porcine cattle head in Galicia, 1986

Another essential innovation has been the opening up of the rural space to non-agrarian activities, such as small and medium sized companies, new types of tourism, and an intense residential growth, due to sub-urbanization processes (Ferrás 1996; Rodríguez González, 1997). This is not to say that nowadays Galician rural space is characterized by a diverse economic structure. Quite contrarily, it continues to be primarily agrarian, with an economy based upon cattle breeding and forestry. However, it is slowly but steadily spreading towards other sectors. Thus, a large number of small towns (ranging from 3,000 to 10,0000 inhabitants), such as Lalín, Verín, Ordes, etc. evidence a certain degree of industrial development. Industrial activities usually revolve around manufacturing centres built on small family businesses, which have expanded thanks to profit investment. Textile and food industries, as well as ironwork construction, are other important sectors. These industries, which usually require a lot of labour, are able to reduce costs due to their coexistence with part-time agriculture (Rodríguez González, 1997).

Another non-agrarian activity, rural tourism, has been promoted by a great number of public policies but, except for some specific areas (Way of Santiago, surroundings of Lugo and northern Pontevedra), the impact of this type of tourism on the economy is still very small. It is interesting to note, however, that its development has changed our perception of the countryside. Our society is now for the first time able to see rural areas as places of leisure, instead of solely the background of the former generations (Santos, 1997). Thirdly, the emergence of sub-urbanization in the outskirts of the cities may cause, on the one hand, the withdrawal of the rural world, due to a diffuse residential growth and, on the other, the disorganization of certain areas where brand new houses are starting to appear alongside with traditional country cottages, meadows, vegetable gardens and industries (Ferrás, 1996).

Principal indicators of the rural world decline in a context of improving living standards

Any analysis of reality is likely to lead us to different conclusions depending on our critical perspective. Thus, during the last forty years, Galician rural society has witnessed the disappearance of food, clothing and housing problems. Indicators reveal that the income of the inhabitants of small villages is similar to that of Galician city dwellers, that is, slightly more than 60% of the European Community average (IGE 1998). This assessment does not include the profit generated within the family unit in terms of subsistence agriculture, informal lease of houses or land, property transactions, etc (Lois and Santos

1995). Today Galician rural population has a steady income; they are mass consumers as any other European, and, as the constant implementation of productions shows, they have been able to adjust to an increasingly regulated and competitive market (regarding milk, meat, wine, wood, etc). Families of farmers, cattle breeders or forest owners have particularly benefited from this situation. Nevertheless, the rural world in Galicia has lost its hegemony to the unstoppable spreading of urbanization. The demographic rates in agricultural areas continue to decline dramatically, and the population is ages at a fast pace. Traditional social networks dissolve whereas there is an increase in mobility and communication.

The inherent paradox is that rural areas die while they experience an economic success they had never imagined before. The unwanted consequences of this success are the abandonment of the land (while some plots are made to produce beyond their means), and the gradual transformation of 75% of the Galician soil into a new periphery, almost a *natural park*. Within this periphery, dozens of thousands of crop lands combine with alternative tourist activities and with some profitable small industries.

Demographic data are, without any doubt, the best indicators of the crisis in the rural areas. Thus, since 1960, the two least urbanized cities in the inland (Lugo and Ourense) have lost 20% of their population, whereas the total population of Galicia slightly increased from 2,602,962 inhabitants to 2,743,399 in 1996 (the increase concentrated on the A Coruña-Pontevedra Atlantic citizen axis). In a study conducted some years ago, we concluded that from 1981 to 1991, the territories with more than 80% of agrarian assets lost 23% of their population; the ones between 60-79.9 lost 16.3%; whereas the territories whose agricultural and fishing assets were less than 20% increased 8% (Lois and Rodríguez 1994).

During the 1990s, rural birth rates were less than 6 per thousand, a percentage lower than the death rate, which went beyond 10 per thousand, with a tendency to increase due to demographic ageing. In fact, according to the 1991 census, 21% of the population in the whole Galician inland were older than 64 (González and Somoza 1997). This situation shows both the devastating effect of the mass migration in the 1960s and 1970s and the threat depopulation in a great number of areas. From a different perspective, it also shows the advantages of the widespread retirement pensions for the familiar financing of the agrarian sector (the Social Security System grants retirement pensions to approximately a quarter of the rural population).

The migration overseas of virtually whole generations, demographic ageing and low birth rates, account for the irreversible decline of peasant communities, which are now reduced to the minimum, whose internal structure has been disturbed. Furthermore, whereas the traditional intensive multiple crop system required social interaction between neighbours, the new models of commercial specialization only involve business contacts with outsiders. Today, mobility is as important in the countryside as in the city. Thirty years ago, rural social networks grouped in parishes, functioning almost as perfect self-centred microcosms; nowadays, the community boundaries are redrawn everyday and there is a tendency for the ties among neighbours to disappear. Nevertheless, some ancestral traditions are still preserved (especially the festive cycle), reinforcing the cohesion within the social group. This group is now a heterogeneous one, consisting of competitive farmers, old peasants living on retirement pensions, who are reluctant to disappear, urban relatives who periodically return to their place of origin , young students, and a large number of distinct groupings of people.

Proposals of territorial balance

So far, this essay has focused primarily on the social and economic aspects of change in the Galician rural world. As a geographer, I am also interested in touching upon the consequences of the change in terms of space and ecology. I have already pointed out that agrarian productivity increased in Galicia by resorting to fewer and fewer pieces of land, which are, in turn, overused (Lois and Santos, 1995; López Iglesias, 1996). This evidence conflicts with the European Community policies promoting land extension, which have proved unable to erase a tendency towards the progressive fragmentation of rural property in Galicia, where the offspring usually inherit the land owned by their parents. This system of inheritance affects the status of the plots conforming the region: the land inherited by people who stay in the village continues to be intensely exploited, whereas that inherited by absent emigrants is either abandoned or planted with trees, and losing its former agricultural function. The families staying in the countryside should be expected to buy the land which has been abandoned or which is being under-utilized at a reasonable price. However, there is evidence that periods where the hectare was too expensive anywhere in Galicia (late 1970s, and 1980s), have been followed by periods where the land market came to a standstill (nobody was prone to buy or sell). The unwanted result of this is that 55-60% of the country's land is regarded as forest, 25% corresponds to urban areas or under-utilized agrarian areas, and the remaining 15-20% concentrates the whole productive effort (Lois and Santos, 1995).

If a particular scale of the agrarian production units reveals sharp contrasts in the use of soil, a general overview of Galicia shows the same opposition between overcrowded areas, and other areas that are gradually being demographically emptied. In fact, 70-75% of the population and wealth concentrate in a vertical stripe including the cities of Ferrol, A Coruña, Santiago de Compostela, Pontevedra and Vigo, which occupies 20-25 % of the whole territory. This 7,000 km area, known as the *Atlantic urban axis,* suffers from internal organization and saturation problems. These problems could intensify in the near future, if the relations with the Portuguese coast expand the urbanized area, until it becomes an important Southern European resort in terms of citizenship, ports and industries (Comisión Europea, 1994; Lois and Martínez, 1998). Nonetheless, as I have already pointed out, rural regression is a fact, in spite of the existence of isolated cases of demographic and industrial growth in small cities (Lugo, Ourense) and in several towns scattered throughout the territory (Verín, Sarria, Lalín, O Carballiño, Vilalba, etc). These isolated examples of economic growth should be seen against a general background of demographic decline and concentration of production in a decreasing amount of land (Rodríguez González, 1997).

The isolation, relative backwardness and unevenness of Galician rural areas have encouraged the European Community to promote local development projects. Galicia is considered a number one priority region and, therefore, programs like LEADER, PRODER, or INTERREG are very well funded (Figure 4.5). The problem is, however, that instead of mitigating the depopulation, the abandonment of the productive land, or the crisis in agricultural communitarian life, these funds have been somehow misused by the local authorities. Therefore, the Development Plans granted by Brussels should be carried out with a more realistic approach, so that the improvement of rural living standards goes hand in hand with the existence of an internally well organized space.

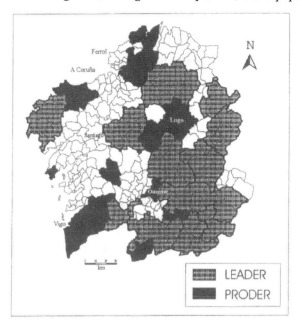

Figure 4.5   Distribution of European programs of rural development in marginal areas of Galicia, 1999

Concluding remarks

As different Galician social scientists point out, the procedures to organize rural space should be the following:

1) the creation of a land bank which allows active farmers to exploit the abandoned land;
2) the establishment of coherent territorial models that are sensitive to regional differences;
3) the encouragement of productive diversification, de-emphasizing rural tourism as the only possible activity, and favouring successful industrial and food enterprises;
4) the privileging of social profit (the settling down of new families, youth employment) over economic profit (which is less important, in spite of what statistics may reflect);
5) the promotion of foreign investment on demographically declining areas by the Public Administration.

References

Beiras, X.M. (1972), *O atraso económico de Galicia*, Editorial Galaxia, Vigo.

Bertrand, J-R. (1992), *A poboación de Galicia. Estudios xeográficos*, Universidade de Santiago, Santiago de Compostela.

Colino, X. and Pérez Touriño, E. (1983), *Economía campesiña e capital. A evolución da agricultura galega 1960-1980*, Editorial Galaxia, Vigo.

Comisión Europea (1994), *Estudio prospectivo de las regiones atlánticas*, Servicio de Publicaciones de la Comisión Europea, Bruselas.

Ferrás, C. (1996), *Cambio rural na Europa Atlántica. Os casos de Irlanda e Galicia (1970-1990)*, Universidade de Santiago-Xunta de Galicia, Santiago de Compostela.

García Fernández, J. (1975), *Organización del espacio y economía rural en la España Atlántica*, Ed. Siglo XXI editores, Madrid.

González, J.M. and Somoza, J. (1997), *O avellentamento demográfico en Galicia e as súas consecuencias*, Documentos de Traballo do IDEGA-Universidade de Santiago, Xeografía III, Santiago de Compostela.

Hernández Borge, J. (1992), *Tres millóns de galegos*, Universidade de Santiago, Biblioteca de Divulgación, Santiago de Compostela.

IGE (1998), *Galicia en cifras, 1998*, Instituto Galego de Estatística-Xunta de Galicia, Santiago de Compostela.

Lois, R. and Martínez, H. (1998), *Xeografía de Galicia*, Editorial Obradoiro-Santillana, Vigo.

Lois, R. and Rodríguez González, R. (1994), 'O retroceso da poboación campesiña e a crise da vida parroquial en Galicia. A importancia da análise microespacial', *Pontenova, núm.* 0, Diputación Provincial de Pontevedra, Pontevedra, pp. 81-93.

Lois, R. and Santos, X.M. (1995), 'As transformacións recentes no campo galego', in *Xeografía Económica I, Xeografía de Galicia, Vol. 5*, Gran Enciclopedia Gallega, Bilbao.

López Iglesias, E. (1996), *Movilidad de la tierra y dinámica de las estructuras agrarias en Galicia*, Ministerio de Agricultura, Pesca y Alimentación, Serie Estudios, Madrid.

López Taboada, J.A. (1996), *La población de Galicia 1860-1981*, Fundación Caixa Galicia, A Coruña.

Méndez, R. and Molinero, F. (eds.) (1993), *Geografía de España*, Editorial Ariel, Barcelona.

Rodríguez González, R. (1997), *La urbanización del espacio rural en Galicia*, Editorial Oikos-Tau, Barcelona.

Santos, X.M. (1997), 'Los caminos de Santiago en la oferta turística gallega', in M. Valenzuela (ed.), *Los turismos de interior. El retorno a la tradición viajera*, Ediciones de la Universidad Autónoma de Madrid, Cantoblanco-Madrid, pp. 571-581.

Torres, Mª.P. de, Lois, R. and Pérez Alberti, A. (1993), *A montaña galega. O home e o medio*, Universidade de Santiago, Biblioteca de Divulgación, Santiago de Compostela.

5   Local poverty in Finland in 1995

JARMO RUSANEN, TOIVO MUILU, ALFRED COLPAERT AND
ARVO NAUKKARINEN

Introduction

Incomes, taxation and income transfers are themes that arouse continuous
interest in the various sectors of our welfare society. The distribution of
incomes and possible imbalances in this attract keen interest among politicians active at different levels in society, among officials and also among
researchers. Income differentials and changes in these in the late 1990s
have also been a major topic in the mass media. Following the declining
economic trends prevailing in Finland in the early 1990s, there has been
even more discussion of the nation being divided into two parts' (e.g.,
comments by the Hunger group and ...; Forma, 1999), into A-class and B-
class, or marginalized citizens (Eva, 1999), or into regions with uneven development.

The present paper analyses three income variables, i.e., the incomes of
household dwelling units subject to state taxation, average disposable incomes and income transfers. The following two factors have greatly affected its content and the questions discussed:

1) Finnish publications and statistics describing economic activity are often
   completely lacking in a regional perspective (e.g. Heikkilä and Uusitalo
   1997; Loikkanen *et al.*, 1998; Statistics Finland, 1998; Uusitalo, 1988;
   Subsistence security survey, 1998), or else the regions are regarded merely
   as an instrument of classification and occupy a marginal position in the
   analysis and presentation of the results.
2) Geographical information systems (GIS), and georeferenced grid cell information in particular, offer new analytical opportunities in terms of technology, methodology and content for use in research into economic activity
   (cf. Haataja, 1995, pp. 11-12).

The present paper endeavours to answer the following two main questions:

1) What are the local and regional differences in income formation and income transfers? and
2) Are there differences between the extremes of settlement, i.e. the most sparsely populated rural areas and the most densely populated parts of cities?

Material and methods

The unit of observation here is the 1 x 1 km grid cell produced by Statistics Finland (cf. Rusanen *et al.*, 1995), applied to material for 1995. In order to protect individual rights, the information was gathered for household dwelling units and does not contain data items referring to less than 5 persons. The coverage of the material is presented by deciles in Table 5.1. Non-response involves the most sparsely-populated areas, for the same reasons of privacy. The population coverage of the material is some 27% and the regional coverage only 12% in Decile 1, i.e. the most sparsely-populated areas, whereas the figures for Decile 2 are 99% and 98.5%. The material covers all cases from Decile 3 onwards. Average population non-response was only 7.5% throughout the material. In practise, interpretations regarding the most sparsely-populated rural areas may involve a degree of uncertainty, but not those for urbanised areas.

The concept of poverty

Defining poverty when examining incomes distribution and measuring it, for example, entails a number of problems (e.g. Chapman and Shucksmith, 1996; de Cunha, 1999; Heikkilä, 1990; Hesselberg, 1995, 1997; Kangas and Ritakallio, 1996, 1999; Ritakallio, 1994; Sommers *et al.* 1998; Townsend, 1993; Uusitalo, 1997), which arise on account of the variables, methods and observation units employed, whether the material was derived from statistics or interviews and whether comparisons involve information collected from different countries on different criteria, etc. It is well known that surveys drawn up from different perspectives may yield heterogeneous results.

The methods employed here comprise basic statistical methods, fractiles and cartographic presentations.

Table 5.1 **Decile limits, population by deciles (*= number of decile), sample sizes and sample percentages**

*	Inhabitants / km²	Popula- tion	Number of inhabited grid cells	Sample Popula- tion	Number of grid cells	Sample % Popula- tion	Inhabited grid cells
1	1 – 19	512,619	82,742	139,198	9,925	27	12
2	20 – 65	500,615	15,531	495,487	15,294	99	99
3	66 – 259	504,627	4,173	504,562	4,172	100	100
4	260 – 534	505,586	1,372	505,586	1,372	100	100
5	535 – 866	503,987	751	503,987	751	100	100
6	867 – 1251	506,783	494	506,783	494	100	100
7	1252 – 1765	503,977	345	503,977	345	100	100
8	1766 – 2569	506,553	243	506,553	243	100	100
9	2570 – 4023	505,893	163	505,893	163	100	100
10	4024 –19120	502,301	83	502,301	83	100	100

Source: Statistics Finland

The poverty limit in the grid cells were determined here by the relative method (e.g. Green, 1996), measured on the basis of the average disposable incomes of household dwelling units in the following manner:

Very poor grid cells
    Average disposable incomes less than 33% of the median incomes in the
    material
Poor grid cells
    Average disposable incomes 34-50% of the median incomes
Almost poor grid cells
    Average disposable incomes 51-70% of the median incomes
Middle-income grid cells
    Average disposable incomes 71-150% of the median incomes
Wealthy grid cells
    Average disposable incomes 151-200% of the median incomes
Rich grid cells
    Average disposable incomes 201-300% of the median incomes
Very rich grid cells
    Average disposable incomes more than 300% of the median incomes.

Income formation and income transfers at the extremes of the regional structure

Fractiles, which include quartiles (quarters), quintiles (fifths) and deciles (tenths) can be used to examine the settlement structure from the most sparsely populated segment to the most densely populated parts of the nuclear centres, regardless of administrative boundaries. Each decile contains an equal number of persons. If the values for the various variables were divided equally between the deciles, no regional differences would exist at all. The differences between the deciles thus indicate deviations from the average.

The decile distributions of all the three variables followed almost the same pattern (Table 5.2). The figures were highest in the third smallest population density decile from which they assumed a falling trend towards the highest and lowest densities. Incomes subject to state taxation were lowest in the most sparsely populated areas and in the largest concentrations of population, i.e. in decile 10. These differences can be considered quite significant, for income levels in the third decile exceeded the national average by 7.6% but were 9% below it in the most sparsely populated rural area and 7.8% below it in the dense population concentrations in general.

Average disposable incomes as percentages of the average income subject to state taxation varied were largest in the second decile (81.5%) and the first decile, falling quite evenly after that, to 76.1% in the most densely populated decile. Thus disposable incomes were higher in the rural areas than in the built-up ones in relative terms. As for average disposable incomes, the distribution differed from the above in that these were lowest in the most densely populated decile.

The dwelling units with the lowest disposable incomes were located in the extremes of settlement. According to a detailed analysis, slightly over a half, i.e. 42 out of the total of 83, inhabited grid cells in the most densely settled Decile 10 are located in the capital, Helsinki, and the rest, with only one exception, in other largest cities. The role of Helsinki at this level can be considered exceptionally prominent.

The above decile analysis points to fluctuations between the income forms within the regional structure. These fluctuations within the deciles are examined in Table 5.3 by means of standard deviations. It can be stated as an evident regularity that variation between the various income forms was largest in the sparsely populated areas, i.e. in Deciles 1 and 2. An

Table 5.2   **Mean incomes subject to state taxation, average disposable incomes and income transfers (FIM) by decile in 1995**

Pop. decile	State-taxable income	Number of grid cells	Disposal income	Number of grid cells	Income transfers	Number of grid cells
1	148,777	9,925	120,742	9,941	13,512	1,587
2	171,287	15,294	139,565	15,295	16,781	12,567
3	178,087	4,172	141,560	4,172	17,033	4,170
4	175,054	1,372	137,681	1,372	16,608	1,372
5	174,042	751	136,021	751	16,518	751
6	174,288	494	135,252	494	16,340	494
7	169,484	345	130,819	345	16,002	345
8	167,720	243	128,676	243	16,054	243
9	160,802	163	123,255	163	16,084	163
10	152,488	83	116,060	83	15,611	83

(1= sparsest, 10= densest decile)

Source: Statistics Finland

Table 5.3   **Standard deviation of incomes by decile (FIM) in 1995**

Population decile	State taxable income	Disposable income	Income transfers
1	79,762	54,930	5,446
2	57,589	39,694	8,927
3	44,502	28,883	5,433
4	45,281	28,887	3,169
5	43,904	28,305	3,004
6	46,968	29,719	3,154
7	45,480	29,112	2,713
8	52,220	31,883	2,982
9	44,090	26,715	3,471
10	28,769	18,275	2,491

(1= sparsest, 10= densest decile)

Source: Statistics Finland

equally obvious trend was that the standard deviation was invariably lowest in the most densely populated areas, i.e. in Decile 10. The smallness of this variation was in a category of its own particularly, as far as taxable and disposable incomes were concerned. The order of income levels varies in Deciles 3-9 without any noticeable regularities. One distinct feature was the quite even distribution of disposable incomes in Deciles 3-7.

The results can be taken to suggest that income levels vary by far the most in the most sparsely populated area, implying that the rural areas contain both extremely prosperous households and ones with very small incomes. The small deviation in the case of population concentrations, i.e. Decile 10, points to equal incomes. The results can also be taken as suggesting that the Finns are still interested in living in rural areas, an option which becomes open to them if their incomes are good. Correspondingly, there would not seem to be very good prospects for increasing income levels to any appreciable extent under conditions which are highly urban by Finnish standards.

The results of the decile analyses are confirmed by the maps given as examples here (Figures 5.1 and 5.2). These are deliberately presented on a scale which should prevent any attempt at identifying individual cells. Local identification elements have also been avoided. The map of the distribution of incomes subject to state taxation in Southern Finland, which contains slightly over a half of the population of Finland, supports the idea that the centres of the major cities in particular stand out on account of their lower average income levels, as in the case of Helsinki, Lahti and Tampere. As for the Oulu region (Figure 5.2), incomes are low in the city centre and the areas close to it.

It can be stated by way of conclusion that, measured using the above meters, the most affluent tenth of the population seem to live in private housing areas which are not very densely populated (Decile 3). It is impossible here to comment on whether Decile 3 represents exactly what the Finns are looking for and aiming at as their ideal living environment, but it is likely that the most wealthy segment of the population are capable of choosing their place of residence more or less according to their own preferences.

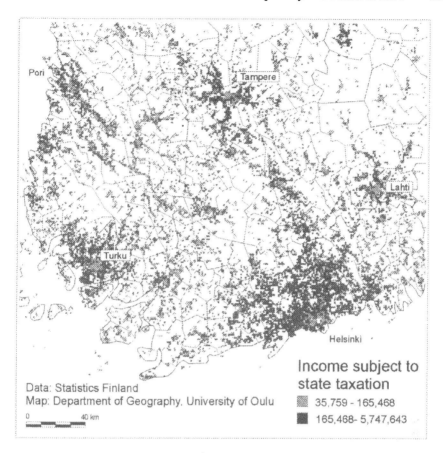

Figure 5.1   Distribution of incomes subject to state taxation in Southern Finland in 1995

Source: Statistics Finland

Effect of distance from a centre on income formation

Taxable incomes, average disposable incomes and changes in these can also be examined with regard to distance from the centre of the administrative district concerned. As no information is available on a point that could be defined as the centre of a local government district, the cell with the

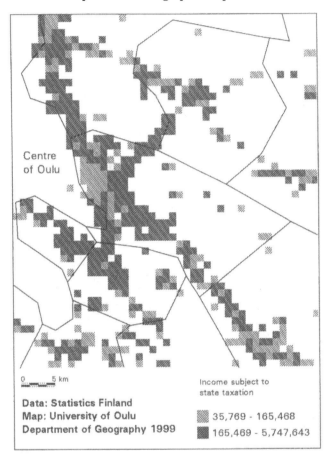

Figure 5.2   Distribution of incomes subject to state taxation in the Oulu region in 1995

Source: Statistics Finland

largest number of inhabitants was taken to represent this centre. In addition, the 10 largest cities were examined separately. Such an approach has been found to be well in line with the idea of where the centre of each local government district is located. The distance was calculated as the crow flies, so that actual travelling distances by road will of course be longer. Coefficients of 1.2 (Martamo and Littow 1992, p. 23) or 1.27 (Lahti and Koski 1993, pp. 72-73) have been employed to express this relation.

Incomes tend to vary greatly between these distance zones (Figure 5.3), usually being very low in and close to the centres and reaching a maximum a few kilometres away. They then tend to fall in an almost linear manner as the distance increases further, apart from areas lying some 15 km or 27 km from the centre, where levels are substantially above those of the surrounding areas. No distinct explanation could be found for these exceptions. Incomes were lowest in the distance zones lying more than 27 km away from the centres. Incomes exceeding the national average were located in the 2-8 km zone in every case except one.

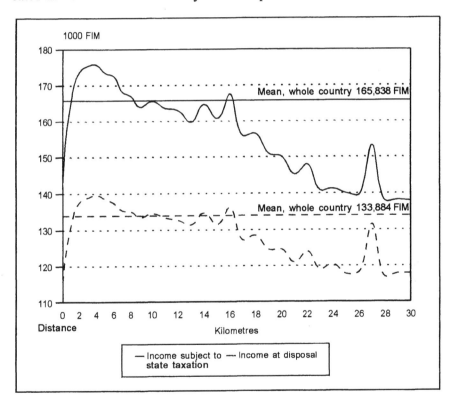

Figure 5.3   **Distribution of taxable and average disposable incomes by distance from the centre of the local government district concerned in 1995**

Source: Statistics Finland

It should be remembered when analysing the results that the analysis concerns all the local government districts in Finland and that there may be quite large differences between them. On the other hand, the results support the conclusions stated earlier. Income levels are the same at the extremes of settlement, although still markedly below those in many other parts of the settlement structure. Although it was stated above that taxable incomes are on the average greater in sparsely populated areas than in built-up areas, those observed in rural areas far away from residential centres are still below the level for the rest of the countryside and of the same magnitude as in district centres.

Poor and wealthy areas in terms of disposable incomes

One topic of major interest in the distribution of incomes concerns the extremes on the scale, which in practise refers to the distribution of the richest and poorest segments of the population and how many of these people there are. The aim here is to examine their distribution between the various parts of the regional structure. Many investigations into differences between countries, for example, classify incomes using the relative poverty method, taking grid cells and their inhabitants as observation units.

The major results are quite clear, as can be assumed on the basis of research into the distribution of incomes in Finland. According to the classification employed here, the majority of the areas, a total of 88.9% (Table 5.4a), and an even larger number of their inhabitants, i.e. 93.6% (Table 5.4b), gained an average income. The absolute numbers of extremely poor and poor people were very small, whereas the proportions of 'almost poor' areas (6.7%) and population (4.0%) suggest an imbalanced income distribution. Deviation between the classes was greatest in the sparsely populated areas, as already indicated in part in the analysis of the deviation in incomes by decile. Thus the sparsely populated areas contain the only extremely poor people, but also extremely rich ones.

The incomes distribution obtained using the decile approach consequently follows the pyramid pattern, the results indicating that the sparsely populated areas contain persons representing all income categories, not all of which can necessarily be found when moving towards the nuclear centres. Decile 10, the most densely populated, contains only people with average incomes or those who are 'almost poor'. Extremely rich people seem to shun densely populated areas in particular.

Table 5.4   **Absolute distribution of grid cells and their population by deciles of average disposable income in 1995**
A) = number of cells, B) = number of inhabitants in cells,
*) = number of decile

Table 5.4a

*	Very poor	Poor	Almost Poor	Middle-income	Wealthy	Rich	Very rich	Total
1	12	226	1,572	7,879	204	35	13	9,941
2	0	11	469	14,123	606	69	17	15,295
3	0	1	84	3,942	127	16	2	4,172
4	0	0	42	1,302	26	1	1	1,372
5	0	0	20	711	19	1	0	751
6	0	0	16	467	9	2	0	494
7	0	0	13	322	9	1	0	345
8	0	0	7	227	7	2	0	243
9	0	0	2	158	3	0	0	163
10	0	0	6	77	0	0	0	83

Table 5.4b

*	Very poor	Poor	Almost poor	Middle Income	Wealthy	Rich	Very rich
1	85	2,072	19,145	114,133	3,206	537	184
2	0	376	13,311	460,519	18,629	2,207	464
3	0	71	10,360	476,331	15,980	1,549	271
4	0	0	14,601	481,103	9,113	428	341
5	0	0	13,429	477,377	12,654	527	0
6	0	0	16,521	479,132	9,326	1,804	0
7	0	0	19,309	470,226	13,109	1,333	0
8	0	0	15,219	472,497	15,091	3,746	0
9	0	0	5,977	491,319	4,587	0	0
10	0	0	48,366	453,935	0	0	0
Tot.	85	2,519	176,238	4,376,512	105,705	12,131	1,260

(1= sparsest, 10= densest decile)

Source: Statistics Finland

One matter of interest from the point of view of social policy was the areas classified as 'almost poor'. More than one fourth of these in terms of population, i.e. 27.4%, were located in the most densely populated decile. Of these six areas, two were in Helsinki and the others in Tampere, Pori, Kuopio and Joensuu. It should be noted, though, that in absolute terms the category 'almost poor' was quite equally represented in the various parts of the regional structure employed here.

According to the National Research and Development Centre for Welfare and Health (Ministry of Social Affairs and Health, 1997), 3.0% of all households and 2.3% of all inhabitants in Finland represented an income category below 50% of the median, the smallest proportion among 13 European countries examined. The figures presented here are slightly higher, but still more or less of the same magnitude. The difference may be attributed to the concept of dwelling unit employed in compiling the material, which is not fully synonymous with the concept of family. The results may also have been affected by the restrictions imposed on collection and organisation of the material by the needs of personal privacy.

Summary and conclusions

It is difficult to measure income distribution and the concepts connected with it, such as poverty and wealth, and different meters easily yield different results. It is also difficult to appreciate the overall situation, too, for hardly any researcher could gain access to the necessary material on account of the sensitivity of the topic alone. The most essential results obtained from the analysis here were the following:

1) Disposable incomes were below the average in the extremes of settlement, i.e. in the most sparsely populated and most densely populated areas. In addition, the incomes levels of the rural areas lying furthest away from the centre of their local government district were below those of other rural areas. The proportion of the 'almost poor' population in the most densely populated areas exceeded the expected figure by a factor of 2.7.

2) Measurement of the income distribution by the relative poverty method indicated that major poverty areas, i.e. cells classified as extremely poor or poor, occurred only in sparsely populated areas. These variables are scarcely relevant for looking for the poorest cells in nuclear centres and suburbs. The population of these cells is small, making up only 0.06% of the total considered here.

3) In addition to the similarity in income levels between the extreme areas, another important observation was that persons classified as receiving an average income evidently make up the largest group when measured in terms of both cells and the number of persons living in these. Finnish welfare policy is thus reflected well in the regional structure.

4) If any conclusions can be drawn on the basis of the high income levels and the place of residence preferences of those Finns who are most able to chose, the message communicated by the present results is evidently that settlements should be sparse rather than dense. This is backed up by the assumption that the most prosperous seek to live in areas which best meet their requirements for a good living environment. The population densities in Decile 3, where both taxable and disposable incomes were largest, varied between 66 and 259 per square kilometre. The results thus suggest that this density is the most attractive. The assumption is also supported by the fact that the richest people live in sparsely populated areas and not in the population agglomerations. It should be noted, however, that the meters employed here do not tell the whole truth regarding income levels, nor do they measure levels of affluence, for example.

Applied to regional structure, the ecological fallacy means here a situation in which the conditions prevailing in some of the areas do not conform to the average set (Martin, 1991). The results obtained here show once again how misleading it would be to rely only on average national or local government data for the purposes of decision-making and planning (cf. also Green, 1995; Kodras, 1997; Lawless and Smith, 1998). It is more a rule than an exception from the point of view of incomes, for example, that income differences within local government districts may be many times greater than those between regions. These differences would seem to be particularly prominent in sparsely populated areas. Thus grid cell data serve to bring out information which would remain obscure or undetected if regional aggregates such as local government districts or NUTS regional units were used.

An ideal material for eliminating the ecological fallacy would comprise co-ordinates defined at the individual level, but this approach is impossible for reasons of both privacy and the availability of material. In addition, material of this kind would increase cartographical responsibility for the way in which the results should be presented at the local level.

The system of urban and rural areas in Finland is still going through a phase of urbanisation accompanied by depopulation of the rural areas, a phenomenon which has already been familiar for decades elsewhere, where it has almost reached a state of balance. The traditional concept of 'poor countryside - rich town' can be assumed to be diversifying in nature, so that unemployment, low income levels, poverty and marginalization will no longer be properties attached to peripheral areas alone in Finland but will be increasingly associated with the structures of the central areas. Unemployment and poverty surveys conducted in Great Britain, for example, have pointed to accumulations of poverty (e.g. Hasluck, 1987; Haughton *et al.*, 1993; Green, 1996). As unemployment has already moved from the rural areas to the towns, it can be assumed that poverty, or most of it, will follow suit. It remains to be seen whether this change will lead to a search for entirely new patterns of social and regional policy solutions to back up the traditional regional policy.

References

Chapman, P. and Shucksmith, M. (1996), 'The experience of poverty and disadvantage in rural Scotland', *Scottish Geographical Magazine*, vol. 112, No 2, pp. 70-75.

de Cunha, A. (1999), 'Urban Poverty on Switzerland: Exclusion Processes and Public Policy', *Swiss Journal of Geography*, Heft 1, 1999, pp. 37-45.

EVA (1999), 'Mielipiteiden sateenkaari, Raportti suomalaisten asenteista', in INTERNET at *http://www.eva.fi./julkaisut/raportit/asenne99*.

Forma, P. (1999), 'Mureneeko solidaarisuus, polarisoituuko yhteiskunta? Hyvä- ja huono-osaisten sosiaaliturvaa ja hyvinvointivaltiota koskevat mielipiteet 1990-luvun Suomessa', *Yhteiskuntapolitiikka* 1/1999, pp. 3-19.

Green, A.E. (1995), 'The Changing Structure, Distribution and Spatial Segregation of the Unemployed and Economically Inactive in Great Britain', *Geoforum*, Vol. 26, No. 4, pp. 373-394.

Green A.E. (1996), 'Changing Local Concentrations of 'Poverty' and 'Affluence' in Britain 1981-1991', *Geography - Journal of the Geographical Association*, vol. 81, part 1, pp. 15-25.

Haataja, A. (1995), 'Tulonsiirtomallit sosiaalipolitiikan arvioinnissa', *Turun yliopisto, sosiaalipolitiikan laitos*, Sarja A:7/1995, p. 104.

Hasluck, C. (1987), *Urban unemployment: local labour markets and employment initiatives*, Longman, London and New York, p. 248.

Haughton, G., Johnson, S., Murphy, L. and Kevin, T. (1993), *Local Geographies of Unemployment*, Avebury, Aldershot, p. 187.

Heikkilä, M. (1990), 'Köyhyys ja huono-osaisuus hyvinvointivaltiossa, Tutkimus köyhyydestä ja hyvinvoinnin puutteiden kasautumisesta Suomessa' (English summary: Poverty and Deprivation an a Welfare State, A Study of Poverty and the Accumulation of Welfare Deficits in Finland), *Sosiaalihallituksen julkaisuja* 8/1990, p. 272.

Heikkilä, M. and Uusitalo, H. (toim.) (1997), 'Leikkausten hinta, Tutkimuksia sosiaaliturvan leikkauksista ja niiden vaikutuksista 1990-luvun Suomessa', *Stakes, sosiaali- ja terveysalan tutkimus- ja kehittämiskeskus, Raportteja* 208, p. 223.

Hesselberg, J. (1995), 'Urban poverty and shelter: an introduction', *Norsk geografisk tidskrift*, Vol. 49, pp. 151-160.

Hesselberg, J. (1997), 'Poverty in the South and in the North', *Geografiska Annaler*, 79B (4), pp. 227-241.

Hunger group and ... (Nälkäryhmän kannanotto ja selvitys suomalaisesta köyhyydestä) (1998), 'Köyhyysongelman ratkaisua hakemassa', *Kirkkopalvelujen julkaisuja* 1, p. 38.

Kangas, O. and Ritakallio, V.-M. (1996), 'Eri menetelmät eri tulokset? Köyhyyden monimuotoisuus', in Olli Kangas and Veli-Matti Ritakallio (toim.), *Kuka on köyhä, Köyhyys 1990-luvun puolivälin Suomessa, Stakes, sosiaali- ja terveysalan tutkimus- ja kehittä-miskeskus, Tutkimuksia* 65, p. 231.

Kangas, O. and Ritakallio, V.-M. (1999), 'Sosiaalipolitiikka vai rakenne, tulonsiirrot, rakennetekijät ja köyhyys Pohjoismaissa ja Ranskassa', *Sosiologia* 1/99, pp. 2-17.

Kodras, J. E. (1997), 'The Changing Map of American Poverty in an Era of Economic Restructuring and Political Realignment', *Economic Geography*, Vol. 73 (1), pp. 67-93.

Lahti, P. and Koski, K. (1993), 'Pääkaupunkiseudun yhdyskuntakustannukset, vaihtoehtoisten rakentamisalueiden vertailu', *Ympäristöministeriö, Kaavoitus- ja rakennusosasto, Selvitys* 5, p. 116.

Lawless, P. and Smith, Y. (1998), 'Poverty, Inequality and Exclusion in the Contemporary City', in Paul Lawless, Ron Martin and Sally Hardy (eds.), *Unemployment and Social Exclusion, Landscapes of Labour Inequality*, Regional Policy and Development 13, Jessica Kingsley Publishers, London and Philadelphia, Regional studies Association, London, p. 280.

Loikkanen, H. A., Rantala, A. and Sullström, R. (1998), 'Regional income differences on Finland, 1966-96', *VATT-discussion papers 181*, Government Institute for Economic Research, Helsinki, p. 52.

Martamo, R. and Littow, P. (1992), 'Kaupunkiseutujen kasvun yhdyskuntakustannukset', *Sisäasiainministeriö, Kunta- ja aluekehitysosasto, Moniste* 7, Huhtikuu 1992, p. 54 + appendices.

Martin, D. (1991), *Geographic information systems and their sosio-economic applications*, Routledge, London and New York, p. 182.

Ministry of Social Affairs and Health (Sosiaali- ja terveysministeriö) (1997), Sosiaaliturva Suomessa 1995, *Sosiaali- ja terveysministeriö, talous- ja suunnitteluosasto* 1997, p. 108.

Ritakallio, V.-M. (1994), 'Köyhyys Suomessa 1981-1990, Tutkimus tulonsiirtojen vaikutuksista' (English summary: Poverty in Finland 1981-1990, A Study of Effects of Income Transfers, *Sosiaali- ja terveysalan tutkimus- ja kehittämiskeskus STAKES, Tutkimuksia* 39, p. 141.

Rusanen, J, Naukkarinen, A., Colpaert, A. and Räisänen, S. (1995), 'Grid-based statistical data for regional research - Finnish experiences', *Nordia Geographical Publications* 24:2, pp. 91-102.

Sommers, L.M., Mehretu, A. and Pigozzi, B.Wm. (1998), 'Rural Poverty and Socio-Economic Disparity in Michigan', in Lennart Andersson and Thomas Blom (eds.), *Sustainability and Development, Out of the Future of Small Society in a Dynamic Economy*, University of Karlstad, pp. 269-277.

Statistics Finland 1995, part 1 (1998), 'Population 1995', *Statistics Finland*, p. 381.

Subsistence security survey 1998 (Toimeentuloturvakatsaus 1998) (1998), Sosiaali- ja terveysministeriö, Kansaneläkelaitos, Sosiaali- ja terveysalan tutkimus- ja kehittämiskeskus, *Sosiaali- ja terveysministeriön julkaisuja* 1998:5, p. 142.

Townsend, P. (1993), *The International Analysis of Poverty*, Harvester, Wheatseaf, p. 291.

Uusitalo, H. (1988), 'Muuttuva Tulonjako, Hyvinvointivaltion ja yhteiskunnan vaikutukset tulonjakoon 1966-1985', *Tilastokeskus, Tutkimuksia* 148, p. 134.

Uusitalo, H. (1997), 'Neljä laman vuotta: mitä on tapahtunut tulonjaossa?', in Matti Heikkilä and Hannu Uusitalo (eds.), Leikkausten hinta, Tutkimuksia sosiaaliturvan leikkauksista ja niiden vaikutuksista 1990-luvun Suomessa, *Stakes, sosiaali- ja terveysalan tutkimus- ja kehittämiskeskus, Raportteja* 208, p. 223.

6 Regional imbalances in Spain at the end of the century

ROSER MAJORAL AND DOLORES SÁNCHEZ-AGUILERA

Introduction

In a previous article (Sánchez-Aguilera and Majoral, 2000), we analyzed the demographic factors used in defining marginal areas in Spain. There we spoke of how the economic changes of the second half of this century had led to a redistribution of population and a far-reaching structural change. This we synthesized in a map entitled 'Marginal Spain. Demographic Factors'. Here, we return to the evolution of the Spanish economy and its regional differences – inequalities which policies implemented to date are failing to resolve.

We present now an analysis of these regional differences as illustrated by economic indicators. In the first part we analyze levels of disposable family income per capita, drawing on a recently published source: *The Commercial Yearbook of Spain*, published by a banking institution.[1] Complementing these statistics, and given the spatial correlation with income levels, we also analyze the characteristics and spatial distribution of unemployment. The possibility of using and mapping information at the scale of the municipalities allows us to analyze features usually dealt with more broadly in most studies of the subject.

In the second part of this paper an analysis is undertaken of two extreme, yet highly representative, situations of two highly contrasting Spains: the prosperous region of the Ebro Axis, in the north of the country and the undynamic region of Extremadura, bordering Portugal in the southwest.

Socioeconomic inequalities in Spain: income levels and unemployment

The distribution of disposable family income (DFI) or economic level is an indicator of a society's wealth and poverty, the inequalities that exist in that society and also the level of social well-being. The spatial distribution of income in Spain is characterized by marked regional contrasts; while unemployment is closely linked to this unequal distribution of income. Indeed, unemployment levels are particularly high in certain Spanish regions and are among the highest in the EU.

Economic level

In general, Spain is among those countries with the lowest levels of wealth in the EU. According to EUROSTAT, in 1980, Spain stood at 71% of the EU average, though in 1991 this had risen to 80%. Only the Balearic Islands recorded levels above the EU average, Madrid and Navarra coincided with the average, while the autonomous communities of Andalucia, Extremadura and Galicia recorded levels below 65%. The income distribution reveals varying degrees of development linked to environmental, social, economic or a combination of some or all of these factors (Paz-Bañez, 1997). Wealth became concentrated in certain communities, which led throughout the sixties and early seventies to the gradual abandonment of the least favourable areas and the concentration of population in certain communities and, in particular, in urban centres (Sánchez-Aguilera and Majoral, 2000). Two distinct areas can be identified in Spain according to income levels: that of the North East, the Levante (Valencia) and Madrid which is characterized by economic levels above the Spanish average and that of the central, western and southern areas of the Peninsula, where the levels are below the average and reach their lowest levels in Extremadura and Andalucia.

At the start of the 1970s, the communities with the highest income levels were the Basque Country, La Rioja, Aragón, Navarra, Catalonia, the Balearics, Valencia and Madrid which, with 48.5% of the population, accounted for 58.4% of DFI. The economic crisis and the policy implemented during this decade saw a worsening of this situation by 1981, when the richest regions – now with 50.73% of the population – accounted for 61.73% of the income. Income redistribution policies implemented by the Socialist Government (1982-1996) managed to reduce this proportion to 57.4% for a similar share of population. However, these policies were not enough to eliminate the imbalances, and in 1998 the two Spains can still be identified, as is shown by the data describing the economic level [2] equivalent to the estimated DFI per capita index for 1997 [3] (Table 6.1).

Table 6.1   Basic economic indicators, Spain (1996-1997)

Region	Population 1996	Unemployment rate 1997	Economic level	Tourist index	Industrial activities
Andalucía	7,234,873	31.75	3	15,372	63,355
Aragón	1,187,546	14.01	6	2,092	22,080
Asturias	1,087,885	21.26	6	1,214	11,559
Balears (Illes)	760,379	11.75	7	15,962	17,266
Canarias	1,606,534	19.87	5	17,605	14,388
Cantabria	527,437	20.88	5	1,061	7,065
Castilla y León	2,508,496	19.45	5	2,925	39,444
Cast-La Mancha	1,712,529	18.60	4	1,234	30,625
Cataluña	6,090,040	17.09	7	15,831	128,082
Com.Valenciana	4,009,329	20.25	5	8,401	66,063
Extremadura	1,070,244	29.24	3	738	12,267
Galicia	2,742,622	18.41	4	3,106	37,236
Madrid	5,022,289	18.36	6	10,241	60,309
Murcia	1,097,249	19.49	4	1,232	14,625
Navarra	520,574	9.97	7	487	11,905
País Vasco	2,098,055	19.08	7	2,054	38,695
Rioja (La)	264,941	11.51	6	292	6,012
SPAIN	39,669,394	20.82	5	100,000	581,841

Sources: La Caixa: Anuario Comercial de España, 1999 and INE: España en cifras, 1998

The average DFI level per capita in Spain in 1997 was between 7,570 and 8,435 €, i.e. level 5. Levels in the autonomous communities range from 3 in Andalucia and Extremadura to 7 in the Balearics, Catalonia, Navarra and the Basque Country. In the provinces, levels range from 2 in Cadiz and Badajoz to 7 in many of the provinces in the north-east of the Peninsula and the Balearics. At the municipal level, there are extreme situations: municipalities with an economic level of 1 in many provinces in the south and some with levels of 8-10 in the wealthier provinces (Figure 6.1).

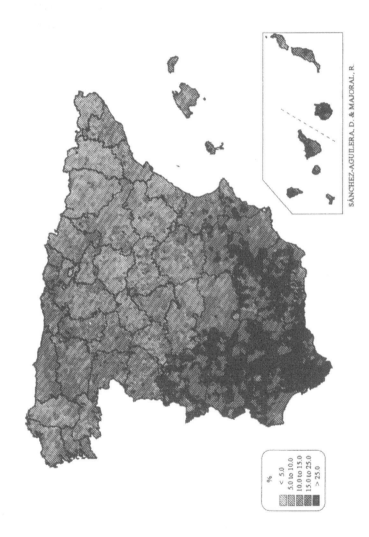

Figure 6.1 Unemployment rate; Spain, 1991

Source: INE: Censo de la Población de España, 1991

Ninety percent of the population (91% of the municipalities with more than 1,000 inhabitants) has an economic level between 3 and 7 (6,000 and 10,000 €, approximately). Five percent of the population (8% of the municipalities) has a level of 1 (less than 5,425 €) or 2 (from 5,425 to 6,000) and the remaining 5% (1% of the municipalities) reach levels 8, 9 and 10, that is they have a DFI per capita greater than 10,000 €. The distribution differs greatly according to the province and community. So, while in Badajoz (Extremadura) 48% of the population has a low economic level – levels 1 or 2 – and in Cadiz (Andalucia), a quarter of the population has an economic level of 1 and none of its municipalities rises above level 3, in Navarra 100% of the population has levels above 5 and 40% has levels of 9 and 10 and in Lleida (Catalonia), 47% of the population has levels of 9 and Madrid, the Balearics and the four communities in the North East (Basque Country, Navarra, Aragón, Catalonia) constitute the prosperous Spain; while Extremadura, Andalucia, Galicia, Castilla La Mancha, Murcia and the Canary Islands find themselves at the opposite end of the spectrum. In an intermediate position, in a kind of transition zone, are those communities in contact with the two Spains (Castilla León, Valencia, La Rioja, Cantabria and Asturias). Income therefore falls from north to south, but also from east to west. Further analysis shows a regionalization based on the dynamics of the Spanish economy and the main axes of development which agrees with most studies conducted in this field (Vallés, ed. 1997):

1) The Cantabrian Cornice with a declining industrial sector and a marked fall in the region's gross domestic product per capita between 1973 and 1991;

2) The Ebro Valley typifies a development axis with sustained production growth and a DFI above the Spanish average (La Rioja is 15 % above);

3) The Mediterranean coast has DFI levels more than 10%;

4) The island groups -dependent on tourism- have a similar growth rate, though in the Balearics (the highest DFI of all the Spanish communities) it was more marked up to 1985, when the growth rate was actually greater in the Canary;

5) In the central regions – comprising the two Castillas – Castilla-La Mancha is thriving (especially around Ciudad Real and Albacete) while Castilla-León is declining, in particular in its northern provinces. This central area, in spite of having improved its relative standing in terms of DFI, is still some 10% below the Spanish average;

6) The economic level of the south (Andalucia, Extremadura, Murcia) has remained unaltered in recent years.

Unemployment

The severity of the economic crisis unleashed at the beginning of the 1990s emphasized one of the main economic and social problems facing western economies, and one that is particularly severe in Spain: large-scale unemployment. After a period of rapid growth in the second half of the eighties, a period was ushered in when many jobs were lost, taking unemployment levels to record highs.

High rates of unemployment have many economic, political, social and ethical implications and are the topic of all socioeconomic debate in the political arenas of advanced capitalist countries. Unemployment can no longer be defined solely as an objective social condition (that of being deprived of a position of work), it is also a subjective condition (experienced differently by each individual) and a recognized civil status (complete with institutions and regulations). In the EU, the convergence processes applied by the member economies – oriented by the aims of the Single Market and the Monetary Union – are causing grave problems, and highlight the need to redefine the European project in order to face the high rates of unemployment and the challenges of international changes. Similarly, it is necessary to bear in mind the deregulation of the job markets in the capitalist economies in the 1990s which, it goes without saying, has an obvious effect on Spain (Paz-Bañez, 1997).

Spain has one of the lowest rates of employment in the OEDC -only half the population between 16 and 64 years of age are active in the job market. This is due, in good measure, to the low levels of participation of women, as male activity rates are similar to those in other European countries. Thus, while the activity rate is around 63% for men, this figured is almost halved for women (37.8%). At a regional level there is a close relationship between activity levels and the incorporation of women on to the job market: the provinces with low levels (Zamora, Córdoba, Cuenca, Guadalajara) usually register lower rates of activity among women. Together with this factor, other variables need to be considered such as demographic ageing, which has obvious implications for the structure of the active population and the processes of industrial conversion, which led to the early retirement of many workers and the incidence of which is marked in certain areas (Asturias, León).

A further feature of employment in Spain is the low degree of job security. A high percentage of work is temporary -the highest in Europe- as a consequence of the reforms made to work contracts in the 1980s and at the beginning of the 1990s. Temporary work is concentrated mainly in certain groups of the population: those joining the job market for the first time, those with the lowest levels of education and women. In terms of sectors, temporary contracts primarily affect those working in agriculture, construction and tourism. Recently, steps have been taken in an attempt to raise the number of permanent contracts, but their effect has been limited.

These measures also seek to solve one of the biggest problems facing Spanish society, unemployment. The progress made in reducing the number of jobless in recent years has been slow and certainly inadequate so as to put an end to the problem. Although recent trends are beginning to be turned round, unemployment continues to be primarily a phenomenon affecting women. Moreover, more than 50% of the unemployed are long-term sufferers, that is they have been without work for more than a year. Seen regionally, the rates of unemployment vary greatly from one community to another. These differences reflect the regional nature of the job market, along with regional differences regarding the branch of economic activity in which people are engaged. Similarly, other variables to be borne in mind are the training levels of the jobless and the proportion of long-term unemployed.

The structure of employment by sector is of great importance. Although in some communities, such as Extremadura and Galicia, the agrarian sector continues to be significant, in all the other communities agriculture is clearly in decline. Moreover, the recent years have been marked by a gradual de-industrialization, in common with other European countries: Asturias, the Basque Country and Catalonia, traditionally the most heavily industrialized of the autonomous communities, are those which have been hit hardest. In terms of the relative standing of the communities there has been considerable stability -those in the south, Andalucia, Extremadura and the Canary Islands- have always had the highest rates of unemployment, while Aragón, the Balearics, La Rioja and Navarra have always been at the other extreme with the lowest levels.

The rate of unemployment still shows a basically north-south trend according to statistics for 1997 (Figure 6.2): the highest rates are in Andalucia (32.7) and Extremadura (27.3) and the lowest in Navarra (9.4), the Balearics (10.4), La Rioja (10.9). An analysis of the provinces shows unemployment is at its lowest point in Lleida (8.1) and at its highest in Cadiz (38.6).

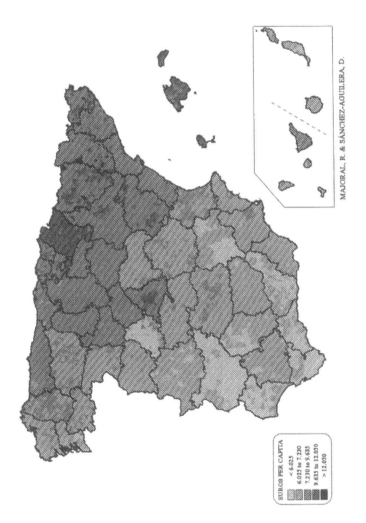

Figure 6.2   Economic level; Spain, 1997

Source: La Caixa: Anuario Comercial de España, 1999

Regional imbalances: two images of Spain

The only regions not to be included in objective number 1 of the European structural funds [4] are Aragón, Catalonia, Madrid, La Rioja, Navarra and the Basque Country. All the other regions were named, in terms of income, as beneficiaries of these funds. This classification establishes, from the outset, a division of Spain which coincides with the situation we have described above and which can be observed in the supporting maps.

Below, we analyze two regions which are representative of these extreme situations: on the one hand, the Ebro Axis, one of the most economically dynamic regions, and on the other, Extremadura which, because of its extreme situation as its name suggests, finds itself at the opposite end of the spectrum (Figure 6.3).

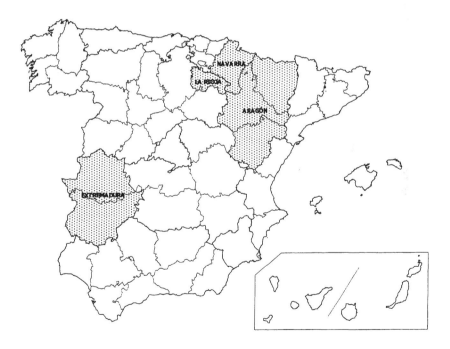

Figure 6.3   The location of the studied communities

The Ebro Axis, a region for the future

Since the 1970s, certain corridors or axes of development have been described which join up, in a Y-shape, the three, traditionally, most dynamic regions in Spain: the two industrial regions par excellence, Catalonia and the Basque Country and Madrid, the capital of the State (Majoral, Andreoli and Cravidao, 1998). Recently, they have acquired even greater prominence due to the rapid development of the axes that link the three nuclei, with marked improvements in their communication infrastructure and a high rate of economic growth thanks to the strengthening of their industrial base.

The Ebro Axis, from Cantabria to the river delta in Catalonia, has become one of the most economically dynamic regions in Spain. Three autonomous communities occupy the central part of its valley: Navarra, La Rioja and Aragón. It is an area of diversified growth and heavy investment which links up two main industrial centres of the Peninsula: the Basque Country and Catalonia. The adjoining regions have a similar, complementary economic profile and are highly attractive to investors. These communities have per capita income levels considerably above the Spanish average, and near the European average, and, as mentioned, unemployment rates that are relatively low and indicators of social well-being above the Spanish average.

The communications network infrastructure is an important element in the development of the Corridor. Thanks to the A2 motorway and A-68 highway, which links the Cantabrian with the Mediterranean communications system, the Ebro valley has been brought closer to the other major axis of dynamism in the Spanish economy: the corridor which runs parallel to the coast from the French frontier to Valencia passing through Barcelona. The Ebro valley, is set to improve its communications in the future, mainly with the building of the high speed train between Madrid and Barcelona and its link with the French frontier passing through Zaragoza (see Font and Majoral, 1998).

The co-operation and trade relations between the three communities are, intense. The wine companies in Navarra buy their grapes from the vineyards of La Rioja. The car component factories in La Rioja supply the Volkswagen factory in Pamplona. The asparagus growers in Aragón supply the food canning plants in Navarra and the people of Navarra shop in the hypermarkets in Zaragoza. These interconnections are also found in agreements involving the provision of health care and other services.

One of the characteristics shared by the three regions is the diverse nature of their economic base in both agricultural and industrial sectors. In 1993, Navarra and La Rioja topped the league table with the highest levels of employment in the industrial sector and are currently competing to be the communities with the highest industrial added value in Spain, higher even than the Basque Country and Catalonia. A large part of recent foreign investment in the Ebro Axis has been in the car and car component industries, linked to two major plants built, primarily, to produce cars for the export market and which are governed by business strategies drawn up far from the Ebro Axis (Burns and White, 1998). Both companies set up business in Spain, at either end of the Ebro corridor, at a time when currency devaluation were frequent - they are the Volkswagen factory in Pamplona and the GM Opel plant near Zaragoza. The plants' suppliers are located along the axis in a corridor which is about 20 km wide, with a high degree of accessibility at a 'just in time' distance from the Volkswagen and Nissan plants in Pamplona, Mercedes Benz in Vitoria and Opel in Zaragoza.

Diversification is, nevertheless, a priority in the central valley of the Ebro and the agro-industries play an important role in this respect. They are a large group of companies, some leaders in their sector, that have grown up thanks to the wealth of the agriculture. The Ebro and its tributaries (originating in the Pyrenees and the Iberian System) irrigate a large part of the three communities (Navarra, La Rioja and Aragón). The city of Haro where some of the main wineries of the *appellation contrôlée* of Rioja are to be found is there. The vineyards and associated industries also have a strong presence in Navarra and recently have begun to expand into Aragón, where the wines of Somontano and Campo de Borja are made in wineries boasting the latest technology in viticulture practices. The aim is to produce quality wines for the top end of the market. Given the size of these viticulture firms, both private firms and co-operatives are merging in order to face the problems of restructuring, which will require the incorporation of better technology and the need to be able to face greater competition.

Rapid growth has also been seen in the production of high quality market garden produce, in particular asparagus and mushrooms. The dynamism of agro-industry is visible in various initiatives. The producers of mushrooms in the area around Calahorra formed the Champiunion co-operative for canning their produce in 1992. Another initiative that benefits the three communities is the setting up of a research centre for the production of asparagus in Tudela (Navarra), the main centre of asparagus firms. The leader in the field is Ian, the largest cannery of asparagus in the world.

In addition to firms working in the automobile and agro-industrial sectors, other companies of note include the Swedish company, Electrolux which has concentrated its production of refrigerators for the European market in La Rioja, and the industry producing sports shoes, mainly for export, sited in Arnedo. In all the industrial sub-sectors in the Ebro Axis there are small and medium-sized companies belonging to larger parent companies, even multinationals. In Navarra alone there are more than 100 companies that are totally or partially owned by foreign companies. In La Rioja also the presence of foreign capital is widespread with firms such as Schweppes, Heinz, Nobel Industries, Sheller, Globe, Draftex and 13% of jobs are in firms with foreign share capital (Burns and White, 1998).

The three communities have invested considerable efforts in the building of industrial parks to attract firms to the region. The park of Fuenteciega near Haro; El Sequero, near Logroño, where Tabacalera, the largest tobacco company in Spain, has its largest and most modern centre; and Tejerias, near Calahorra, are just some examples. In Navarra, industry has located in the Alsasua-Pamplona-Tudela corridor and is characterized by its integration in the international markets as an exporter of manufactured goods and an importer of primary materials, semi-manufactured goods and machine equipment.

The industrial area of the central valley of the Ebro is highly competitive in the Europe of the Euro, with a rate of investment higher than in the rest of Spain. Investment is still helped by a favourable exchange rate, the incentives offered by the local authorities, and a workforce which is still cheap. However, in the Europe of a single currency it will face the challenge of remaining competitive by maintaining its high rate of productivity.

Extremadura: a region seeking economic modernization

At the opposite end of the economic spectrum lies Extremadura, which as its name indicates occupies a place on the periphery. It is a region that has lived for many years anchored to its past and where the changes that are brought with economic modernization -such as the introduction of new economic activities and the development of transport infrastructure- have been accepted with difficulty. This has been exacerbated by its peripheral location as regards the centres and axes of the greatest economic dynamism in the country.

The transport infrastructure remains, even today, greatly inadequate in his region. Its frontier character and the distance separating it from the main development axes account for the poor transport and communications networks. The three main corridors of transport cutting through Extremadura (the N-V which connects Madrid with Lisbon and which is currently being made into a dual carriageway; the N-630 which crosses the Peninsula from Sevilla to Gijon and the N-430 which links Badajoz with Valencia) contribute to the regional organisation but they are clearly insufficient when dealing with increased interregional flows (particularly the N-630 and the N-430).

These comments on the dearth of road communications can be extended to a range of other communications: first, to the railway, in which the only two lines of note are the international routes that link Madrid with Portugal; second, to air transport, centred on the only airport in Badajoz which does not have the facilities to handle the transport of goods, and finally, to the number of telephone and telex lines which sees Extremadura occupying the last place among the Spanish regions.

This deficiency in infrastructures has for a long time been accompanied by an economic marginality. Extremadura is not only one of the least favoured areas in Spain but in the whole of the European Union. Extremadura is one of the areas making the lowest contributions to the Spanish gross added value (estimated at 1.9%). In addition, statistics from the mid-1990s show that the GDP per capita is the lowest in Spain, equal to 46% of the EU average. Moreover, if regional data of the distribution of disposable family income per capita are compared with the Spanish average, major imbalances become apparent: nearly 50% of the population of Extremadura has income levels below a million pesetas while this percentage is considerably lower in Spain as a whole (5%); moreover, all the municipalities of Extremadura have an income level below 7,530 €, an income threshold which corresponds to just 37.5% of the Spanish population. However, the economy has improved as a consequence of transfers from the State and the EU, increasing disposable income by 20%.

These low income levels are in part explained by the characteristics of the job market in Extremadura. Activity rates are low because of demographic ageing and because of the low rates of employment among women.

What's more, Extremadura, together with Andalucia, has the highest rates of unemployment in Spain. With rates around 30% of the active population, unemployment in Extremadura is a structural phenomenon that derives from its productive structure and a poorly trained and largely unqualified workforce. The main features of its economy are its unremitting backwardness and the serious limitations in its structure: a stagnant agrarian sector with low rates of productivity, a weak and imbalanced industrial sector and services which are overinflated and lethargic (Zapata, 1996, pp. 654).

The imbalances in the productive system are the result of a long historical process. Since the 19th century specialization in the agrarian sector has been strengthened through the expansion of the cultivated area and the introduction of technical innovations which permitted a greater use of farm lands. Throughout the 20th century there has been a gradual substitution of work by capital accompanying the improvement in reservoirs and the channelling of water (extension of irrigation in what was the pioneering *Plan Badajoz*) and the introduction of new crops such as corn, sunflowers, tomatoes and tobacco.

Both in the case of Caceres and Badajoz, the agrarian sector continues to be of great importance, a relatively high percentage of the population still works in agrarian activities (18.4% in 1995). Moreover, a high degree of agrarian specialization can be observed given that the contribution of the agrarian GAV is much higher than the region's average.

Extremadura is typically described as a non-industrialized region or marginal in terms of industry (Bosque and Méndez, 1995). Clearly, this has been the sector with the weakest growth in the regional economy. As a sector it is poorly developed and engages only a small percentage of the active population, less than half the Spanish average, and is characterized by the number of very small firms (80% employ fewer than five people). Similarly this region is specialized in a series of sub-sectors (energy and water, food and drinks, wood, cork and furniture, textiles and building materials) which generate a low level of added value. Moreover, they are intensive in the nature of their work.

Industry in Extremadura is subject, in large part, to a marked dependency on other areas. The energy sub-sector produces six times more energy than it consumes and satisfies the needs of other central regions, which exploit its low population density to locate unpopular power centres, such as the nuclear plants of Almaraz and Valdecaballeros. Moreover, both the agro-industry and the cork industry in Extremadura specialize in the initial stages of the transformation process. Industrial firms tend to locate in nuclei with more than 10,000 inhabitants and along the main axes of communication, in particular the N-630 and the N-V. The main centres are the provincial capitals and that of the autonomous community – Mérida – where the diversification of this sector is greatest. Zafra, where the machinery plant of Deutz Diter is located, Don Benito and Plasencia are other industrial centres of some regional importance.

Within the tertiary sector Extremadura is beginning to exploit its tourist potential. The region has a variety of attractive tourist resources – primarily its artistic heritage from the Roman and Baroque periods and the beauty of its natural landscapes – that are currently under-exploited. Moreover, and unlike the Ebro Axis, which has developed a good infrastructure as regards rural tourism, initiatives in this type of tourism in Extremadura are slow in coming to fruition and it has yet to play a significant part in the regional economy.

Concluding remarks

Regional imbalances still form part of the picture of modern-day Spain. These economic differences have been inherited from disparate processes of economic modernization, both temporally and spatially, and have had the effect of strengthening the peripheral nature of certain areas. The policies implemented in the period of development – the 60s and the beginning of the 70s – were not only unable to reduce regional differences but actually widened the gap. In the last two decades, the policies implemented by the Spanish authorities – at the level of the state and the autonomous communities – and by the EU have aimed, in the main, to reduce the differences in income between the regions. However, the effectiveness of these policies has been limited. This is exemplified by the fact that, despite receiving the highest level of state investment in relation to its gross domestic product, Extremadura continues to be the poorest region in Spain. In this region, which between 1985 and 1997 received the equivalent of 8.4% of its regional GDP, the negative consequences of its low level of economic integration have been virtually unaffected.

Spain today is characterized by the persistence of marked contrasts in income levels – the North East shows higher levels of income, while the South West suffers higher levels of unemployment, to the extent that it has become a structural problem against which it is difficult to apply effective measures.

In contrast, certain areas such as the Ebro Axis stand out in terms of their dynamism, benefiting from their strategic position – between two of the main industrial poles of the country – and an excellent transport infrastructure. This has given rise, in these regions, to real economic diversification which as well as adding to their dynamism constitute, without any doubt, one of their main assets in order to maintain their position when facing the challenges for the Spanish economies of integration within a Europe with a single currency.

Notes

1. *The Commercial Yearbook of Spain* is a statistical compendium of data related to the municipalities, grouped by province and autonomous communities, detailing the economic and commercial characteristics of Spain. Of the 8,098 municipalities in Spain, the Yearbook publishes data for the 3,196 with more than 1,000 inhabitants. These municipalities, which account for 96% of the population, comprise, however, only 62.7% of Spanish territory.
2. Economic level is a synthetic indicator of the average level of disposable income of the inhabitants of a municipality. It is the income available for spending and saving, or what amounts to the same thing, the sum of all income received in a given period, so it can be considered as the total wage plus interest on savings, social benefits and transfers, less direct taxation paid by the family and Social Security payments. Information is drawn from the regional records of the National Institute of Statistics and it is calculated using 20 variables of disposable Family Income per inhabitant. The index was calculated by the Klein Institute of the Universidad Autónoma of Madrid and published in the *Commercial Yearbook of Spain*.
3. *The Commercial Yearbook of Spain* defines 10 income levels: 1. less than 5,425 €; 2. 5,425-6,025; 3. 6,025-6,775; 4. 6,775-7,570; 5. 7,570-8,435; 6. 8,435-9,635; 7. 9,635-10,845; 8. 10,845-12,050; 9. 12,050-13,250 and 10. more than 13,250 €.
4. In application of article 8.2 Regulation (EEC) n°2081/93 of the Council (20 July, 1993) as stated in appendix 1.

References

Bosque, J. and Méndez, R. eds. (1995), *Cambio industrial y desarrollo regional en España*, Oikos-Tau, Vilassar de Mar.

Burns, T. and White, D. (1998), 'Spains's Ebro Valley', *Financial Times*, Monday, Dec. 7, 1998, pp. 27-29.

Caixa d'Estalvis i Pensions de Barcelona (1999), *Anuario Comercial de España*, Servei d'Estudis de la Caixa, Barcelona.

Font, J. and Majoral, R. (1998), 'MAGLEV trains in the Iberian Peninsula: Some considerations and proposals', *Transrapid-Verkher in Europa, Geocolleg*, 11, pp. 51-66, Gebrüder Borntraeger, Berlin-Stuttgart.

Majoral, R., Andreoli, M. and Cravidão, F. (1998), 'Regional perceptions of marginality. A view from Southern Europe', in H. Jussila, W. Leimgruber and R. Majoral (eds.), *Perceptions of Marginality. Theoretical issues and regional perceptions of marginality in geographical space*, Ashgate Publishing Ltd, Aldershot, pp. 147-164.

Ministerio de Economía y Hacienda (1994), *Plan de Desarrollo Regional 1994-1999*, Madrid.

Paz Bañez, M.A. (1997), 'La distribución de la renta' in Vallés, J. (coord.), *Economía española*, McGraw-Hill, Madrid, pp. 343-367.

Sánchez-Aguilera, D. and Majoral, R. (2000), 'Demographic factors contributing to regional imbalances in Spain', in H. Jussila and R. Majoral (eds.), *Globalization and marginalization in geographical space*, Ashgate Publishing Ltd., Aldershot, pp. 210-225.

Vallés, J. coord. (1997), *Economía española*, McGraw-Hill, Madrid.

Zapata, S., ed. (1996), *La industria de una región no industrializada: Extremadura, 1750-1990*, Universidad de Extremadura, Cáceres.

Internet sources

http://www.cfnavarra.es
http://www.larioja.org
http://www.juntaex.es
http://www. aragob.es

Acknowledgements

This paper has been prepared as part of a research project entitled *Delimitación y análisis de las áreas marginales en Cataluña*, funded by the Dirección General de Investigación Científica y Técnica (DGICYT) of the Ministerio de Educación y Cultura (Research Project: PB95-0905), and by an *Ajut de Suport a la Recerca dels Grups Consolidats del II Pla de Recerca de la Generalitat de Catalunya* (1997SGR-00331).

7 Job development and regional structure – local examples of growing and declining industries in Northern Finland

TOIVO MUILU

Introduction

Marginal areas are often lumped together as one entity in public discussions of the vicious circle that results from a fall in the number of jobs and active population in certain areas. Marginality as such is a controversial concept which difficult to define (Leimgruber 1993). It is taken here to apply to local government districts (municipalities) with a fairly small population and located far away from the major central areas of Finland. Discussions of this problem at the national level have for a long time been based on the opposition between the declining regions of Eastern and Northern Finland and the developing regions of Southern Finland and the coast. Monitoring of unemployment, which was the major topic that attracted attention in the recession years of the 1990s, has revealed regional and local differentiation even between rural areas, however, of a kind that cannot be explained by 'traditional' ways of thinking.

The main trend in regional structure in the late 1990s, when Finland has been recovering from the recession, has been the strengthening of the position of the largest centres. These are usually university towns characterized by high population growth and the rise of new jobs, mainly in the high-tech sector and related services. There are only a few growth centres of this kind in Finland, and only two lying outside Southern Finland, of which the most prominent is the Oulu sub-region. Many provincial and municipal centres are suffering from population loss (SVT, Väestö, 1998, p. 11), while the sparsely populated areas have been losers in the overall process of regional development, a trend which has been noted in a number of economic policy surveys (e.g. Holappa and Haveri, 1999; Valtakari, 1999; YTR-Raportti 5/1999).

The attraction of growth centres thus seems to be concentrating Finland's population within certain areas, which of course tends to lead to recession in the marginal areas. The purpose of the present paper is not to rediscuss the already well-known negative trends, but to point out that there are municipalities even in the declining countryside of Northern Finland where trends in population and jobs have been more favourable than in neighbouring areas, on account of either particularly sensible targeting of development resources or some other local factor. In such a case, one could even talk about local success stories. The present paper compares Sievi and Suomussalmi, two rural municipalities in Northern Finland which are both characterized by a fairly small population and a somewhat remote location but overtly different development trends. These will be considered both on the basis of statistical time series and contextual information from www. Both municipalities are considered here to be marginal, even though Sievi is located close to the coast while Suomussalmi is a more typical remote locality in Northern Finland. They will be compared here with each other, with other rural-type municipalities in Finland and with the overall area in which they are located, i.e. the province of Oulu. Finally, an attempt will be made to point to possible causes of the different trends observed in them.

Basic facts on Sievi and Suomussalmi

Some basic facts on Sievi and Suomussalmi are presented in Table 7.1. The municipality of Sievi is located on the southern border of the province of Oulu and of Northern Finland, and has an area of some 782 km^2. Its birth rate is among the highest in Finland, and the population which can be considered youthful, in that the proportion of persons aged under 20 years is greater than elsewhere in the country and that of the age-group under 16 years one the highest. Its population has expanded steadily throughout the 1990s. Sievi has been classified as an EU Objective 5b area (development and structural adjustment of rural areas). Its centre is located just beside main road no. 28, some 25-40 km away from the centres of the neighbouring municipalities (see also http://www.sievi.fi; http://europa.eu.int/comm/dg05/esf/en/public/overview/chap7.htm).

Table 7.1   Some basic facts on the municipalities of Sievi and Suomussalmi in 1997

	Sievi	Suomussalmi
Land area (km^2)	782.3	5,275.5
Population, total in 1997	4,969	11,692
0-14-year-olds (%)	27.5	17.0
15-64-year-olds (%)	58.1	66.0
65- year-olds (%)	14.4	17.0
Excess of births	38	-12
Total net migration	22	-205
Economic dependency ratio	2.0	2.4
Industrial structure		
Agriculture and forestry (%)	29.4	18.1
Manufacturing (%)	32.6	21.0
Services (%)	36.3	56.6
Road distance to Oulu (km)		
(capital of the province)	146	221
EU Objective region (see text)	5b	6

Source: Statistical Yearbook of Finland 1998, pp. 76-77

Located in Kainuu, Suomussalmi has some of the oldest ancient dwelling sites in Finland, dating back almost 10,000 years. It has 151 km of joint border with Russia and is the eighth largest municipality in Finland in area (5,275 km^2). It is also extremely sparsely populated (2.1 inhabitants per square kilometre), so that it has been classified as an EU Objective 6 area (development and structural adjustment of regions with an extremely low population density). Approximately a half of its population live in built-up centres. The low population density is well reflected by the fact that the length of the public roads in Suomussalmi is the second highest in Finland (1,072 km) (see http://www.suomussalmi.fi/; http://europa.eu.int/comm/dg05/esf/en/public/overview/chap7.htm).

The population of Sievi is thus younger and its economic structure more dominated by primary production. In Suomussalmi, the birth-rate and migration balance are negative and the economic dependency ratio (ratio between passive and active population) evidently below that of Sievi. Differences in development between the two municipalities are also suggested by their demographic trends (Figure 7.1).

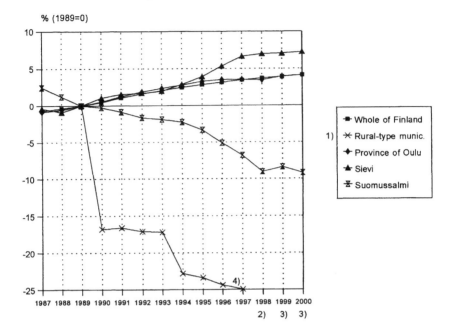

Figure 7.1   Population development in the study areas in 1987-2000

Notes: 1) Rural-type municipalities refers to the classification proposed by Statistics Finland in 1989. There are two definitions for rural-type municipalities in Finland (Kunnat 1996: 12):
  (a) less than 60% of the population is living in built-up areas, with the
  largest centre having a population of less than 15,000; or
  (b) at least 60% but less than 90% of the population is living in built-up
  areas, with the largest centre having a population of less than 4,000.
2) Estimated population.
3) Population projections including migration by Statistics Finland.
4) Population data for rural-type municipalities from 1987-1988, estimate
for 1998 and projection for 2000 are missing.

Sources: Statistical Yearbook of Finland 1998, SVT Työmarkkinat 1992:14, SVT
Väestö 1989:4, 1990:11, 1993:15, 1995:2, 1996:5, 1997:3, 1997:15,
1998:2, 1998:6, 1998:11, 1999:3, 1999:4, SVT Väestölaskenta 1990, Osa
1

The population of Finland has increased slowly since the late 1980s, as
has that of the province of Oulu, while the rural-type municipalities (see
explanation in the legend to Figure 7.1) have shown a sharp fall in popula-

tion. The population of Sievi is equivalent to less than a half of that of Suomussalmi, but it has increased rapidly throughout the period examined here and is expected to do so in the foreseeable future, while a further decline is forecasted for Suomussalmi. According to the most recent population data, for the first quarter of 1999, Sievi is expanding even more rapidly than expected, its population already having exceeded 5,000 (5,047), whereas an unexpectedly dramatic fall has occurred in that of Suomussalmi, to 11,272 (SVT, Väestö 1999, pp. 27-28).

The two municipalities differ even more dramatically in their trends in unemployment (Figure 7.2), the rate of which increased markedly in all the reference areas in the recession years of the early 1990s, the pattern being more or less the same in the country as a whole, in the rural-type municipalities and in the province of Oulu. At the recovery stage, however, after the most severe unemployment year of 1993, Sievi and Suomussalmi have developed in entirely opposite directions, for where the unemployment rate in Suomussalmi has been among the highest for a long time, remaining permanently above 30% even at the recovery stage, the figure for Sievi is 'only' slightly over 10%, i.e. considerably below the figures for the reference areas, and it is still falling.

What is the reason for the difference in demographic and unemployment trends between Sievi and Suomussalmi? This will be examined below by comparing the various branches of the economy in order to indicate those that account for the what has even been referred to in recent years as the 'Sievi miracle' or 'improbable success story' (e.g. Kangasharju *et al.*, 1999, p. 47).

Development in industries

The local authority employment statistics published by Statistics Finland since 1987 provide information on trends in the employed labour force in the light of a classification of the industries into 9 branches (SVT, Väestö, 1997, p. 15). Trends in these separate branches in Sievi and Suomussalmi over the period 1987-1996 are shown with reference to the situation on the last day of each year in Figures 7.3a-b. Mining and quarrying and electricity, gas and water supplies are combined with the category manufacturing on account of the small numbers of people employed in these.

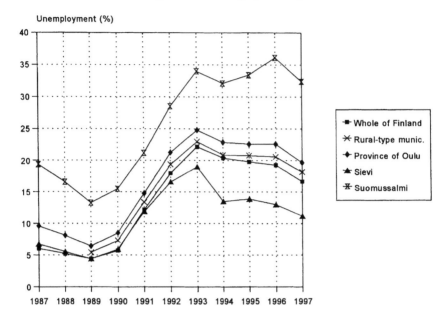

Figure 7.2   Unemployment rates in the study areas in 1987-1997

Note: Unemployment data for rural-type municipalities from 1987-1988 is
                 missing. For definition of rural-type municipalities, see Figure 1.

Sources: SVT Työmarkkinat 1992:14, SVT Väestö 1989:4, 1990:11, 1993:15,
                 1995:2, 1996:5, 1997:3, 1997:15, 1999:3, SVT Väestölaskenta 1990, Osa
                 1

The trends shown in Figures 7.3a and 7.3b point very clearly to a continuous fall in the number of jobs in Suomussalmi, and alarming situation
as it means that one cannot actually speak of the economic recession having come to an end there or of innovations having spread to the area, at
least not by the end of 1996. There were more than 5,000 jobs at most in
Suomussalmi in 1989, and the number had fallen by almost 1,700 by the
end of 1996. The number has decreased in all industries, and at a rate even
faster than the decline in population, which largely accounts for the rise in
unemployment rate and its persistently high levels. Suomussalmi is also
characterized by a high proportion of public sector jobs, mostly in the social services (Figure 7.3b).

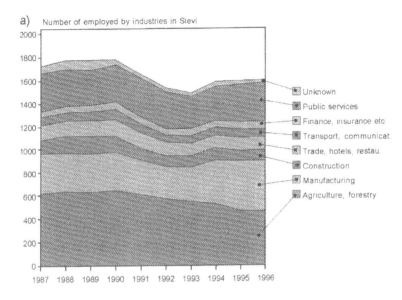

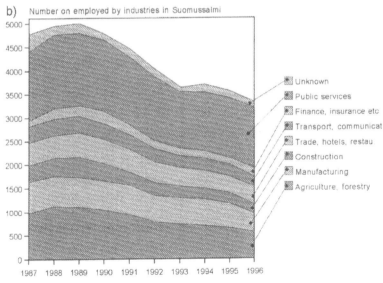

Figure 7.3   Employed labour force resident in (a) Sievi and (b) Suomussalmi by industries in 1987-1996

Sources: SVT Työmarkkinat 1992:14, SVT Väestö 1989:4, 1990:11, 1993:15, 1995:2, 1996:5, 1997:3, 1997:15, 1999:3, SVT Väestölaskenta 1990, Osa 1

In Sievi the role of primary production is still prominent relative to the situation in Suomussalmi, although the proportion of labour force engaged in agriculture and forestry has continued to fall in both municipalities. Their development differences are reflected particularly in the trends observable after 1993, the peak unemployment year, in that Sievi has been able to recover from the recession and to increase its number of jobs so that there were only 177 less at the end of 1996 than in 1990, the most prosperous year economically. The private sector, and manufacturing in particular, has been active in creating new jobs. Sievi is not as dependent on public sector employment paid for from tax revenues as is Suomussalmi, although the number of jobs in this branch has increased slightly in both municipalities relative to the situation that prevailed following the public sector expenditure cuts of the recession years. This growth may be largely attributable to an increase in the number of persons hired through employment support measures (Figures 7.3a and 7.3b).

Employment trends in the total employed labour force, and in the three principal industries, i.e. agriculture and forestry, manufacturing, and public service, in Sievi and Suomussalmi are compared in Figures 7.4 and 7.5 with the average situation in all rural-type municipalities in the country and with that in the province of Oulu. The three industries concerned covered more than two thirds of the jobs in Sievi and Suomussalmi. The year 1989 was selected as the reference year, for two reasons: the national rate of unemployment was lowest at that time, and a statistical classification covering rural-type municipalities was adopted by Statistics Finland in that year, so that information on these is not available for the years 1987 and 1988.

The employed labour force living in rural-type municipalities fell by as much as 40% in the seven years in 1989-1996. Perhaps quite a surprising finding was that all the northern reference areas succeeded better in this respect, in that the fall was not so prominent there, despite the fact that Suomussalmi came close to the average recorded for such municipalities over the whole country. As stated above, only a minor fall occurred in the number of jobs in Sievi, the trend being an upward one by the end of 1996. It should be borne in mind when assessing the generally favourable picture for the province of Oulu, however, that most of the new jobs were created in the city of Oulu itself (Figure 7.4b).

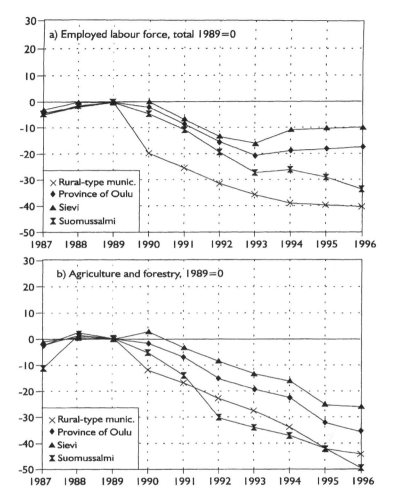

Figure 7.4   Employed labour force resident in study areas in 1987-1996: a) employed labour force; b) agriculture and forestry

Sources: SVT Työmarkkinat 1992:14, SVT Väestö 1989:4, 1990:11, 1993:15, 1995:2, 1996:5, 1997:3, 1997:15, 1999:3, SVT Väestölaskenta 1990, Osa 1

Sievi has also been the most successful municipality in terms of the number of agricultural and forestry jobs remaining, although the trend even here is an obvious declining one. Dairy farming in particular is still very

important in the area. In Suomussalmi the number of jobs in primary production has dropped by as much as half, partly due to a fall in the number
of forestry jobs but also in response to the difficulties encountered in agriculture, as most of the local farms are small. It should be noted, however,
that the number of such jobs has fallen rapidly throughout the 1990s, i.e.
even before the actual recession years and Finland's membership of the EU
(in 1995), which the agricultural organizations in particular considered to
be the principal threat to agriculture in Finland in those years (see Figure
7.4).

The differences between Sievi and Suomussalmi are most prominent
as regards the number of manufacturing jobs, Sievi being in a category of
its own as far as industrial development is concerned. Where the number of
manufacturing jobs in the rural-type municipalities of Finland fell by some
44% in 1989-1996, it increased by as much as 30% in Sievi, despite an obvious fall during the recession years of 1989-1992. The number of industrial jobs in the province of Oulu began to rise again in 1993, but an accel-
erating fall was recorded in Suomussalmi that continued to the end of 1996.
It is thus already evident at this point that a major factor contributing to the
different trends in Sievi and Suomussalmi is the development of manufacturing jobs since the recession year of 1992 (Figure 7.5a).

Suomussalmi is largely dependent on public services (Figure 7.5b),
with the public sector accounting for 38.5% of all jobs at the end of 1996,
as compared with 21.1% in Sievi. The role of public administration jobs in
Finland, particularly in rural areas, increased markedly as a result of the regional policies pursued in the 1970s and 1980s, until they accounted for as
much as 28.4% of all jobs in rural-type municipalities. The public sector
was then cut back drastically in the recession years, which was again reflected in the situation in the rural municipalities, with Suomussalmi as an
extreme example. The turning point in the northern reference areas was
1993, the year of the most severe unemployment, and that at the national
level in 1994, at which point the number of public service jobs again began
to rise. Sievi was also most successful here, in that the number of public
sector jobs there was some 12% larger at the end of 1996 than in 1989. It is
impossible to evaluate the types of new public administration jobs on the
basis of the statistical data employed here, although it may be presumed
that many of them must be temporary ones resulting from the effective labour policy measures adopted in Finland (see Figure 7.5b).

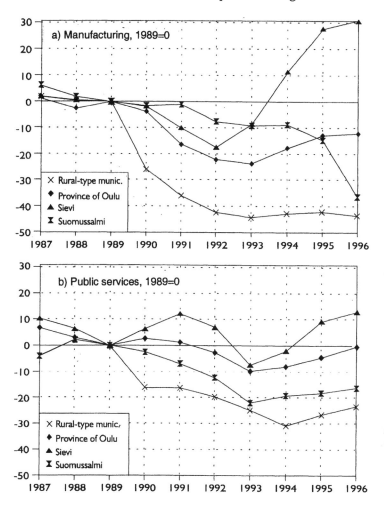

Figure 7.5   Employed labour force resident in study areas in 1987-1996: a) manufacturing; b) public services

Sources: SVT Työmarkkinat 1992:14, SVT Väestö 1989:4, 1990:11, 1993:15, 1995:2, 1996:5, 1997:3, 1997:15, 1999:3, SVT Väestölaskenta 1990, Osa 1

No figures will be presented here as to trends in other industries. In any case, Sievi can be said to have succeeded better in all branches, i.e. construction, commerce, accommodation and catering, transport and telecommunications and finance, insurance and real estate. The situation is poorest in the construction sector, where the number of jobs at the end of 1996 was 40-60% smaller in all the areas examined here relative to the situation in 1989, nor have the other branches listed above usually yet regained their 1989 levels, either.

The 'Sievi miracle' can thus be pinpointed to the secondary sector, the exceptionally favourable trend in which has often earned Sievi a place as a national example in various surveys (e.g. Kangasharju *et al.*, 1999, pp. 38-44; Volk, 1999, p. 113). The possible reasons for the current differences between Sievi and Suomussalmi will be discussed below.

Two municipalities and two paths of development: where do the differences arise?

It is impossible to provide an in-depth discussion of the fundamental causes that have led to the different development trends in the two municipalities, but some conclusions can nevertheless be drawn. These are listed in Table 7.2.

The regional structure of Sievi and its surroundings is largely dominated by agriculture. Sievi and Suomussalmi are both located far away from Oulu or any other major centre, though Sievi's location close to the coast can be said to be more advantageous than that of Suomussalmi. The rate of unemployment in most of Sievi's neighbouring municipalities is below the average for the province, whereas the unemployment rate in the area surrounding Suomussalmi has been the highest for a long time, suffering in particular from the decline in the number of seasonal jobs resulting from the mechanization of timber harvesting. Historical factors and factors connected with regional structure and regional policy have thus undoubtedly played a role in the above differential development, though it is difficult to specify individual factors and their effects in the absence of more extensive analyses.

The question remains as to how a small rural municipality such as Sievi has managed to increase its number of industrial jobs so substantially without compromising on the prominent position of primary production. According to the Sievi home page (http://www.sievi.fi), its industrial development does not rely on one industry only, but rather growth has taken place in the mechanical, electronics and footwear industries.

Table 7.2   Some factors contributing to development differences between Sievi and Suomussalmi

Sievi	Suomussalmi
Location and connections	
Far away from large centres, but appropriate distance to e.g. export harbour	Far away from large centres, remote, sparse population. Could location on Russian border be a possibility?
Population structure	
Young, growing population	Ageing, declining population
Industrial structure	
Versatile and balanced, but primary sector is still important	Based mostly on public sector
Manufacturing	
Several rather big (family) companies, versatile structure, directed to export	Small-size companies
Industrial policy and culture	
Active and innovative, strong support for entrepreneurship, affirmative business atmosphere, long tradition for locally-based activity	Weaker tradition for entrepreneurship, but industrial policy has been activating during recent years

Scanfil Oy, for example, is a leading manufacturer of mechanical parts for the electrical and electronics industry in Finland, employing 600 persons in 1997, although not all in Sievi. Sievin Jalkine, a family company employing 350 persons, is the leading manufacturer of protective footwear in Northern Europe, exporting more than a half of its production. The third major company, Teho Filter Oy, was established as a family enterprise in 1963 and is nowadays one of the main manufacturers of filters in the Nordic countries, with a payroll of 100 persons. Sievi's home page gives a simple explanation for the municipality's success: 'The largest companies in Sievi manufacture their products with the domestic market in mind, although still investing vigorously in exports, from which good results have already been gained. Thus Sievi managed to keep its rate of unemployment below that of the neighbouring municipalities even during the economic recession. Sievi is well-known for having the lowest unemployment rate in Northern Finland' (http://www.sievi.fi/).

The economic image conveyed by Suomussalmi's home page (http://www.suomussalmi.fi) is not as clearly industrial as that of Sievi, focusing instead on the natural environment and history of the area and the opportunities offered by these for tourism and small-scale enterprises. The industrial companies operating in Suomussalmi are usually fairly small, although there are now a number of business development projects based on local initiative and with EU or other public funding taking place to activate small businesses under the themes such as 'Food Kainuu' or 'Nature Kainuu', for example. The economy of Suomussalmi seems to becoming more balanced and the debt situation is well under control (Kaleva 13.4.1999). In addition, some 100 new industrial jobs will be established in the rock processing industry in 1999-2000. Tulikivi Plc has begun to exploit a local steatite deposit (Kaleva 20.4.1999) and EVOX Oy, a manufacturer of electronics and condensers, is concentrating its production in Suomussalmi (Kaleva 27.5.1999). The latter is already the largest industrial employer in the municipality, with a staff of 180 persons (http://www.suomussalmi.fi).

Conclusions

The purpose of this paper was to indicate that development in small rural municipalities located in remote marginal areas does not necessarily have to be negative. The successful economic policy pursued by Sievi is based on a few local family enterprises which survived the recession years most of all by virtue of the flexibility which the locality afforded them and by focusing on exports. Regional policy and local government support for key companies seems to have played an important role in this.

No business based on local conditions has not been set up in Suomussalmi, at least so far, and its economy relies heavily on public sector jobs. It can be assumed that, compared with Sievi, Suomussalmi has suffered more from the changes in branch structure caused by deregulation and privatization, in that it has not been able to replace the lost public sector, primary production and private service jobs with new market-oriented jobs to a sufficient extent. The business policy practiced by Suomussalmi seems to rest on two aspects: on tourism and other small-scale enterprises based on the local natural resources and initiative, and on a few medium-sized enterprises utilizing local natural resources (steatite, timber) or the good supply of local labour. It is estimated that there is also a large teleworking potential in Kainuu (Keränen, 1998). Another aim is to exploit the EU structural policy in a maximally effective manner, and there are also many projects taking place that benefit from location close to the border with Russia.

The most recent local government statistics available for the present purpose were from 1996, so that the development that has taken place since that time cannot be evaluated in a reliable manner here. Preliminary population data and forecasts suggest, however, that Sievi and Suomussalmi are continuing along the above paths of development, although it remains to be seen in the near future whether Suomussalmi will be able to reverse the trend on account of its more balanced economy and more active economic policy. The situation in Sievi in any case serves as an encouraging example 'improbable success story' and of successful co-operation between local intellectual and material resources and local government actors in marginal areas.

References

Holappa, J. and A. Haveri (1999), 'Pohjoisten seutujen elinkeinot, nykytila ja erityisolosuhteiden hyödyntäminen' (English summary), *Acta* nro 103, Suomen kuntaliitto, Helsinki, p. 168.

<http://europa.eu.int/comm/dg05/esf/en/public/overview/chap7.htm> 15.06.1999.

<http://www.sievi.fi/> 15.06.1999.

<http://www.suomussalmi.fi/> 15.06.1999.

Kaleva 13.4.1999 (newspaper article), 'Suomussalmen talous vahvistui'.

Kaleva 20.4.1999 (newspaper article), 'Sata uutta työpaikkaa'.

Kaleva 27.5.1999 (newspaper article), 'Evox Rifa keskittää Suomussalmelle'.

Kangasharju, A., J.-P. Kataja and V. Vihriälä (1999), 'Suomen aluerakenteen viimeaikainen kehitys' (Abstract: Regional development in Finland), *Pellervo Economic Research Institute Working Papers* No 17, Helsinki, Finland, p. 50.

Keränen, H. (1998), 'Työ ja paikka: etätyön kehittymislähtökohdat Kainuussa', *Research Reports 6, Research and Development Centre of Kajaani (REDEC), Regional Development*, Kajaani, p. 125.

Kunnat 1996, 'Alueluokitukset' (Municipalities 1996, Regional classifications), *Handbooks* 28, Statistics Finland, Helsinki, p. 61.

Leimgruber, W. (1993), 'Marginality and marginal regions: problems of definition', in Chang-Yi D. Chang *et al.* (eds.), *Proceedings of the IGU Study Group on Development Issues in Marginal Regions*, Taiwan National University, Taipei, Taiwan, R.O.C, pp. 1-18.

Statistical Yearbook of Finland 1998, Statistics Finland, Helsinki, p. 679.

SVT (=Suomen Virallinen Tilasto) *Työmarkkinat 1992:14* (OSF=Official Statistics of Finland, Labour Market 1992:14), *Työssäkäyntitilasto 1989* (Employment Statistics 1989), Statistics Finland, Helsinki, p. 219.

SVT Väestö 1989:4 (OSF Population 1989:4), *Työssäkäyntitilasto 1987* (Employment Statistics 1987), Statistics Finland, Helsinki, p. 111.

98   Sustainable Development and Geographical Space

SVT Väestö 1990:11 (OSF Population 1990:11), *Työssäkäyntitilasto 1988* (Employment Statistics 1988). Statistics Finland, Helsinki, p. 210.

SVT Väestö 1993:15 (OSF Population 1993:15), *Työssäkäyntitilasto 1991* (Employment Statistics 1991), Statistics Finland, Helsinki, p. 169.

SVT Väestö 1995:2 (OSF Population 1995:2), *Työssäkäyntitilasto 1992-1993* (Employment Statistics 1992-1993), Statistics Finland, Helsinki, p. 273.

SVT Väestö 1996:5 (OSF Population 1996:5), *Työssäkäyntitilasto 1993-1994* (Employment Statistics 1993-1994), Statistics Finland, Helsinki, p. 321.

SVT Väestö 1997:3 (OSF Population 1997:3), *Työssäkäyntitilasto 1994-1995* (Employment Statistics 1994-1995), Statistics Finland, Helsinki, p. 244.

SVT Väestö 1997:15 (OSF Population 1997:15), *Työssäkäyntitilasto 1995-1996* (Employment Statistics 1995-1996), Statistics Finland, Helsinki, p. 379.

SVT Väestö 1998:2 (OSF Population 1998:2), *Väkiluku kunnittain ja suuruusjärjestyksessä 31.12.1997* (Population by municipalities and in order of size in 31.12.1997), Statistics Finland, Helsinki, p. 32.

SVT Väestö 1998:6 (OSF Population 1998:6), *Väestöennuste kunnittain 1998-2030* (Population projection by municipalities in 1998-2030), Statistics Finland, Helsinki, p. 104.

SVT Väestö 1998:11 (OSF Population 1998:11), *Väkilukuarvio kunnittain 31.12.1998* (Population estimates by municipalities in 31.12.1998), Statistics Finland, Helsinki, p. 12.

SVT Väestö 1999:3 (OSF Population 1999:3), *Työssäkäyntitilasto 1996-1997* (Employment Statistics 1996-1997), Statistics Finland, Helsinki, p. 325.

SVT Väestö 1999:4 (OSF Population 1999:4), *Väestön neljännesvuositilasto 1999*, 1. neljännes (Quarterly statistics on population 1999, 1. quarter), Statistics Finland, Helsinki, p. 38.

SVT Väestölaskenta 1990, Osa 1 (OSF Population Census 1990, Volume 1), *Väestön taloudellinen toiminta* 1990 (Economic Activity of the Population 1990), Statistics Finland, Helsinki, p. 321.

Valtakari, M. (1999), 'Maaseutupolitiikka suomalaisessa aluesuunnittelussa', *Helsingin yliopiston maantieteen laitoksen julkaisuja B 46*, Helsinki, p. 160.

Volk, R. (1999), *Muuttuva maaseutu, maaseutupoliittinen selvitysraportti*, Moniste, Helsinki, p. 131.

YTR-raportti 5/1999, *Maaseutupolitiikan yhteistyöryhmän toimintakertomus 1995-1998 sekä kehittämisehdotukset vuosille 1999-2001*, Maaseutupolitiikan yhteistyöryhmä, Helsinki, p. 76.

8 Rural development and marginal areas in Slovenia

STANKO PELC

Rural areas in Slovenia

The dilemma about what is rural and what is urban is not easy to solve. We can always say that rural areas are those that are not urban. Wießner (1999, p. 300) presents different definitions of rural areas. The one given by the state government of Bavaria in 1976 is that rural areas are the areas without the so-called 'Verdichtungs-Räume' (areas of concentration, densely populated areas). Gatzweiler's definition from 1979 cited in the same paper gives the following characteristics of rural areas:

1) Small population density, dispersed settlement pattern and missing efficient (higher level) centres,
2) Peripheral position with regard to the main economic and population centres that is even more obvious because of the lack of adequate connections to long distance traffic network,
3) Widely extended insufficiency of supply and infrastructure that is further reinforced by poor accessibility of quality working places and infrastructure available in the areas of concentration ('Verdichtungs Räume').

These characteristics are not very easy to measure. Even if we accept the simplest definition, we cannot limit the rural areas without knowing which settlements are urban and what their boundaries are. The UN recommendations for Census 2,000 suggest that the settlements with less than 2,000 inhabitants can be part of rural areas while the settlements with 2,000 or more inhabitants can be treated as urban. Haggett (1983, p. 360) states that the size of the settlement considered as urban varies very much from country to country (from a few hundred in Iceland to 20,000 in Netherlands). The notion of a city is multidimensional and complex. Cities have a certain physiognomy, important economic and social functions and they are also a kind of sophisticated technical system. They change through time and we can observe different kinds of cities in different parts of the world. Slovenia, for example, has no big cities and no large urban agglomerations. The number of settlements is almost 6,000 and considering the population of around 2,000,000 living in the area of around 20,000 square kilometres we can easily jump to the conclusion that the majority of these settlements is very small and that they are densely scattered more or less all over Slovenia. The research made by Drozg (1999) for the Slovene government deals with the definition of settlements that meet certain standards necessary to define them as cities. He used physiognomic, economic, demographic and legal criteria. By applying the first three criteria, he could define 48 cities. Some settlements even twice the size of the smallest among those 48 did not match all the three mentioned criteria.

Figure 8.1 shows the population size of these 48 cities in comparison with the idealized rank size curve. In 1869 the correlation between the actual and the expected population size was 0.993 and in 1991 0.978. This change is the consequence of the growth of the capital city of Ljubljana and of the insufficient development of medium size cities (for Slovenia between 20,000 and 100,000 inhabitants). At present, Slovenia has only two cities with more than 100,000 inhabitants and only 3 with more than 20,000 inhabitants. The great majority of Slovene settlements is therefore not urban (but is it really rural?) and even some of those that have adequate population size do not meet the economic, physiognomic or demographic criteria necessary to be treated as cities. Consequently, the share of urban population is very low (41.3% of inhabitants live in 48 cities and 51.5% in settlements that meet the population size criteria of 2 000 inhabitants). On the other hand, a relatively high share of the Slovene population lives an urban way of life outside urban settlements.

If we considered all the settled areas outside urban settlements (48 cities) as rural, these would include small towns (even up to 7,000 inhabitants) and local centres as well as the highly urbanized settlements at the fringes of the largest Slovene cities. In this case, we would have a variety of very different rural (or better not strictly urban) areas from typically suburban areas to peripheral and extremely marginal ones.

On the other hand, we could also define a set of criteria for rural areas and settlements. The meaning of the word rural itself offers the physiognomic criteria - like the countryside (land consisting of woods and fields with few houses or other buildings) (the Oxford Dictionary of Modern English). Some definitions of rural areas emphasize the close connection of

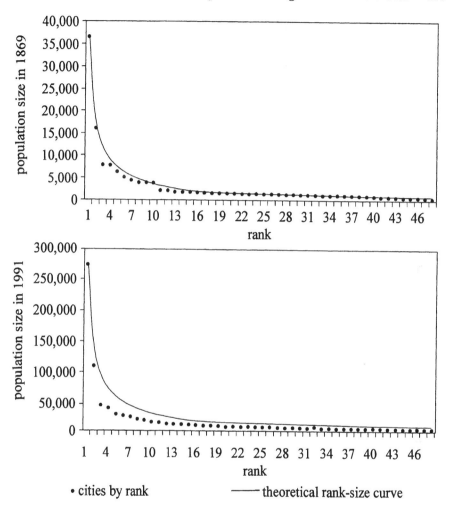

Figure 8.1   **Population rank-size diagram for 48 Slovene cities according to the population size in 1869 and 1991**

these areas with nature and farming as well as having the following characteristics:

1) They are sparsely populated,
2) They have structural problems,
3) They have economic disadvantages and
4) They have experienced poor development in recent decades.

We can thus conclude:

1) Some areas are neither urban nor rural!
2) Marginal rural areas are rural areas with peripheral position and typically rural characteristics!

The developmental policy has to take into account the fact that different types of non-urban areas have very different characteristics and have to be treated in different ways. Each type needs a specific approach to the developmental problems. This means that we have to form different developmental strategies. Wießner (1999) states the following categorization for planning purposes that was designed by the German authorities:

1) Areas with a favourable position with regard to the areas of concentration and interregional and long distance traffic connections,
2) Structurally weak rural areas with very few working places other than agricultural,
3) Areas attractive for tourism at a higher level than regional,
4) Areas with favourable conditions for agriculture,
5) Areas with some concentration and a weak tendency of economic growth that mostly have good accessibility to economic centres.

This categorization is far from being ideal and does not include structurally weak old industrialized rural areas. In Slovenia, these would be the coal mining areas that are already or will be abandoned in the near future. These areas can also experience a certain kind of marginality in the economic as well as in the social sense while typically marginal areas in Slovenia would be the structurally weak ones with peripheral position (close to the state border or at the border of settled land). They certainly need special attention within rural development. Without developmental support, these areas do not have a bright perspective for the future. We can expect their disintegration from the viewpoint of cultural landscape (depopulation, land overgrown with forest - development in the direction of new wilderness areas).

Specific characteristics of rural areas in Slovenia

According to its historical development as well as its natural characteristics, Slovenia is part of Central Europe. As an associate member of EU, it is becoming part of Europe in the political sense too. However, we must not forget the recent past when Slovenia was politically part of the so-called Eastern Bloc. This had some serious consequences not just in the economic development in general but in the development of rural areas as well.

Socialization of land did not affect Slovenia in the same way as it did some other former socialist countries. Farmers did not lose land ownership. However, their ownership was limited to the maximum of 10 ha. For this reason, we can today observe the predominance of extremely small, economically weak farms. In 1993, the average farm was four times smaller than in the 15 EU member countries. In 1991 the share of agricultural population was rather small (7.6%) similar to Finland or Italy and less than Greece, Ireland, Portugal and Spain. In comparison with available farming land, there are almost four times more workers per ha in Slovenia than the average in the 15 EU member countries. On the one hand, this is the consequence of an unfavourable size structure of Slovene farms, and of unfavourable conditions for farming on the other.

Gams (1991) stated poor natural conditions for agriculture as one of the geographical constants of Slovenia. Slovenia has a large share of farming land in hilly regions with very steep slopes, which makes ploughing cultivation difficult. Annual precipitation up to 3,500 mm in the Northwest results in humid to perhumid climate and marshes on the plains. There is also a highly specific run off which causes rapid washing off of nutritious substances from the soil. The acid soils on older Quaternary terraces are therefore another restrictive factor for agriculture. These unfavourable natural conditions together with the fact that Slovenia was from 1919 part of Yugoslavia, resulted in agriculture that had almost no importance for the Slovene economy. Within Yugoslavia, Slovenia was its industrial part while agriculture had great importance in more favourable eastern parts, mainly in Slavonia and Voivodina. The socialist authorities were not favourably inclined towards farmers who represented an undesirable private sector. In such conditions, de-agrarization was quite rapid and rural areas experienced huge socioeconomic restructuring. The result has been the already mentioned low share of agricultural population but a surprisingly high share of households with agricultural holdings (32.3%). Part time farming is therefore very common. Such conditions caused close ties of the Slovene population to the agricultural land. Even today, the market for this land is not expanding. The recent physical growth of Slovene farms is mainly due to the denationalization process and only in very few cases to purchases (more often to lease).

The positive characteristics of Slovene rural areas are the still present spirit of solidarity and reciprocity. Modern farming technology has replaced the earlier co-operative way of farming but the socialist period brought about a great need for co-operation in the field of infrastructure. Rural areas could improve their infrastructure only by engaging their own resources. Many roads, electrical power lines, waterworks etc. were built with the participation of the local population in material, money and work. In socialist times such activities were highly desirable and helped to preserve the willingness for the realization of common tasks so important for the development of rural areas.

Integral rural development and village renewal

In Slovenia the project of integral rural development and village renewal started in 1991. It followed the Bavarian example. The project was lead by the Ministry of Agriculture, Forestry and Nutrition and had very modest funding. Very few officials within the Ministry were working on the project full time.

The project was based on the bottom-up approach and had several small local projects. These small projects were financed partly by the Ministry and partly by the municipalities. The role of the Ministry was one of co-ordinative one while the executive role was handed over to a competent (planning) institute chosen by the municipality based on a public tender. The competence of these institutes, established by the state during the socialist period and some new privately owned ones with mostly only a few employees was not at a very high level. In most cases, the task was completely new for them and often they could not manage to solve the problems they faced in a proper way.

The local projects were supposed to have three stages:

1) The preparatory stage (approx. one year);
2) The introductory stage (approx. one year);
3) The implementation stage (a long term task).

The lack of funding left many expectations of the localities included in the project unfulfilled. In many cases, the preparatory stage was not followed up by the introductory stage. Even more critical was the fact that many localities, which received the funding for the introductory stage, did not successfully complete it and the activities in the following stage were not carried out. The basic characteristics of the mentioned three stages are:

1) The preparatory stage is performed by the initiative group within the locality that wants to include in the project and to receive public funding for the development of their locality (a village or a local community). In many cases the municipal planners chose the locality and stimulated the local population. The final result of this stage was widely spread information about the project, readiness of the local population to become involved in the project and the application for a public tender of the Ministry for the funding of local projects. Some 15 or 20 localities are chosen for funding each year.

2) The introduction stage performed by a competent institute includes the following main tasks:
   • To educate and train the local population for developmental thinking and acting;
   • To formulate the developmental philosophy of the locality;
   • To find local developmental motivation ('spiritus movens' of the locality);
   • To elaborate the synthetic developmental plan of the locality on the basis of SWOT analysis including objectives, tasks and the list of developmental projects that are to be implemented in the following stage;
   • To encourage the active participation of the local population;
   • To form a local developmental initiative group capable of launching developmental projects.

3) The implementation stage is based on the public funding of selected common projects. The localities apply for funding based on a public tender of different Ministries for projects in different fields (infrastructure, small business, agriculture, culture, social affairs etc.). There are also funds available for private investors to get favourable loans.

In recent years the project of integral rural development and village renewal has shown its strengths in the strong local initiative and a highly motivated local population prepared for collaboration and joint action in attaining common goals. However, we must stress that there are also many drawbacks. One is certainly the lack of sufficient funds in the implementation stage. The number of local projects is too high; therefore, only few projects get the necessary public financial support. There is also a lack of adequately qualified professionals necessary for successfully managing developmental projects. Another important weakness is the fact that there are not enough rural development projects at the regional and none at the national level. Consequently, there is too little professional assistance to the local initiative. Some problems that should be solved at the regional and national level are therefore also more or less successfully solved in a similar way in different localities. The absence of a rural development plan and, even worse, the absence of a strategy of social development make the wider success of rural integral development impossible.

To make a step in the direction of more effective rural development Slovenia has to elaborate a *Rural development plan*, which is also a necessary condition for the EU funding of Slovene rural development projects. Some basic elements it should consist of are:

1) A national strategy for the development of rural areas;
2) Common tasks - national level projects;
3) Guidelines for regional and local development projects.

The main goals of rural development in Slovenia are similar to those in other parts of the world, among which the most general one is a higher standard of living in rural areas. This should be achieved by:

1) New (non-agricultural) working places in rural areas;
2) Farm tourism (countryside);
3) Home crafts as an additional source of income;
4) Other activities of the rural population: (intellectual) services, production etc.;
5) Enlarged added value on the farms;
6) Improved rural infrastructure;
7) Reinforced social contacts and the lack of connectivity of the rural society;
8) Strengthened local cultural identity;
9) Re-established sustainability of the rural areas etc.

In 1994, the importance of agriculture for the national GDP was still well above the EU average, in Slovenia, however, the share of agriculture in GDP was at the time lower than in Greece and Ireland. According to the latest data of the Statistical Office of the Republic Slovenia, the share of agriculture, hunting and forestry in national GDP for 1997 was only 3.7%. We expect a further decrease in the importance of agriculture for national economy. On the other hand, we must not neglect the fact that agriculture plays a very important role in preserving the cultural landscape. In the future this may become an important source of income because the urban population will sooner or later and in one way or another have to pay for the scenery of the rural areas they want to enjoy during weekends and holidays. However, farmers must not become gardeners only and rural areas cannot change into a huge park and recreation area. Agricultural production will have to preserve its important productive function even in unfavourable natural conditions, a fact that the developers of Slovene rural areas will have to take into consideration.

Characteristics of marginal rural areas in Slovenia

The marginal rural areas in Slovenia can be defined as peripheral by their position with regard to the lower regional as well as more important local centres. The consequences of this peripheral position are both economic and social. Based on demographic data we can say that the settlements constituting these areas are located in:

1) The hilly area between Idrija and Tolmin (*Idrijsko-Cerkljansko hribovje*),
2) The north-eastern part of Karst with the neighbouring low hills in the region of Vipavsko,
3) The hinterland of the Koper coastal region,
4) The Sv. Vid hills and the *Bloke* plateau south of Ljubljana,
5) Hilly area between Litija, Trebnje and Sevnica,
6) *Bizeljsko* and *Kozjansko* on the eastern border with Croatia,
7) *Haloze* (particularly the Western part),
8) Parts of *Slovenske gorice* (particularly the south-eastern part),
9) Eastern and northern part of *Prekmurje*,
10) The fringes of settled land in other highlands border and mountainous regions.

In Figure 8.2 the location of the areas from the above paragraph (lines 1-9) is shown with ellipses and for the areas from the last line (10) with polygons. It is shown schematically and is more or less approximate. From the demographic point of view the majority of settlements in the listed areas is facing depopulation (local centres are often an exception). In the eighties this depopulation was already the consequence of unfavourable age structure (high share of old population - ageing index over 100) than the consequence of high out migration. Consequently, these settlements have very low birth rates and very high death rates, therefore the natural increase is very low or (which is more common) negative (decrease for more than 1%).

In the period of industrialization young people migrated out of these areas to distant cities where they could get a job. Others just moved to

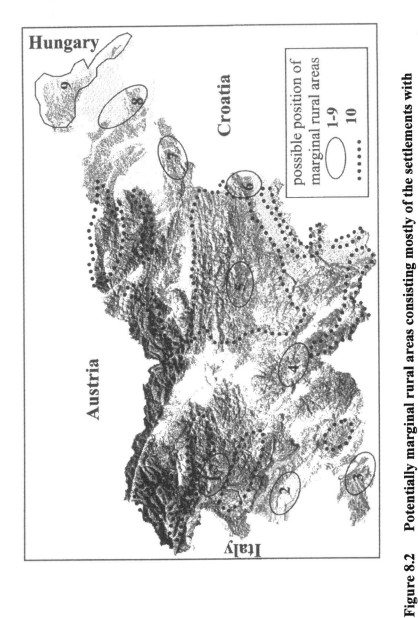

Figure 8.2   Potentially marginal rural areas consisting mostly of the settlements with unfavourable demographic conditions (from 1 to 10 according to the above text)

smaller local centres in the valleys that had better accessibility to working places and services. In recent decades fast growing motorization (365 passenger cars per 1,000 inhabitants in 1996) improved the accessibility in many rural areas and opened them to intense urbanization impacts. This caused strong polarization of space from suburban and highly urbanized settlements in the close vicinity of the cities on the one hand through partly urbanized and stable rural settlement to endangered and decaying settlements in marginal rural areas on the other. If we travel through the Slovene countryside along the motorways, we can hardly observe any sign of marginality in the physiognomy of the cultural landscape. This is only possible in more remote areas and in the areas with a higher relative altitude. The most common and obvious sign is no doubt the grasslands overgrown with forests. Other characteristics in the landscape that are not as obvious are:

1) Bad accessibility to working places and to services in local and regional centres,
2) Unsatisfactory supply in the area and the lack of services,
3) Infrastructural deficiencies:
   - A lack of good roads (the dispersed settlement pattern with low population densities and poor or hardly suitable natural conditions for road construction),
   - Insufficient water supply network,
   - Other (electricity, sewage, phone, information…),
4) Low income and therefore poor economic standard of the local population mostly employed in agriculture and in the secondary sector.

Market oriented agricultural production is not very common in these areas and self-sufficient farms are not a rare exception. They are quite common and usually owned by older farmers without any successors. Agricultural productivity is very low mainly because of unfavourable natural conditions that make mechanized cultivation impossible and also for some other reasons such as the size of the farms which are too small farms and have many small plots etc. People that are only part time farmers invest the money they earn with their non-agricultural employment into their farms. For that reason, many farms are over mechanized. On the other hand, the market for agricultural land is still not very active. People do lease their land but they do not want to sell it unless necessary. However, that is more a problem in the areas where there is enough flat land suitable for cultivation. In hilly areas, there are often no farmers that would want to buy or even take land into lease. Cultivation is simply not profitable enough for that.

Conclusions

I would like to conclude this paper with some thoughts about the threats and perspectives of marginal rural regions in Slovenia.

The threats for these areas are:

1) The continuation of negative trends leading to depopulation and socioeconomic decay,
2) As a single region within an enlarged EU - Slovenia will not meet the criteria for Objective 1 (regions which are lagging behind economically, with a GDP less than 75% of the EU average),
3) The lack of public funding for the development of these areas,
4) Too small young active population necessary for the realization of the developmental tasks.

The continuation of present trends threatens the rural marginal regions to become a depopulated wilderness. In the final stage of depopulation that will cause many social problems as well as a great loss of resources that will not be easily re-activated, not to mention the disintegration of very attractive cultural landscape.

If Slovenia is to be treated as a single region within the enlarged EU, there is a great fear that it would not meet the criteria for Objective 1 and therefore the necessary funding of the economic development from that source will not be available. This also means that there will be further lack of finances for the development of Slovene rural areas, particularly those that are marginal. With GDP per capita of 9,163 USD in 1997 and an annual real growth of 3.9% in 1998, Slovenia will not be able to provide enough funds for efficient developmental support for the majority of its marginal rural areas. It is not very likely that this development will get a priority in the national (developmental) policy.

Another very serious threat is the unfavourable demographic situation in rural marginal areas, which will make the development of these areas impossible. There can be no development without human resources. In addition, the former experiences show that it is simply not possible just to 'import' these resources. Without young people living in the area the possibilities for the success of new developmental projects are very small or none.

Negative perspectives of marginal rural areas in Slovenia can be derived from the presumption that there will not be enough funds available for the developmental support, therefore Slovenia will invest only in promising marginal rural areas that have more resources of their own. That means that only the areas with better demographic conditions (higher share of younger population) and with more natural and cultural resources would get the necessary developmental support. These areas could expect the preservation of agricultural production and cultural landscape while the rest of the marginal rural areas would disintegrate, which would lead to expanding wilderness areas and to a shrinking cultural landscape. This would also cause the already mentioned growing of social problems particularly for the older inhabitants of the disintegrating areas, such as unequal possibilities etc.

Positive perspectives are to be tried out by using the opportunities offered by the development of new information technologies. With investments into the basic information infrastructure, new possibilities for dispersed working places can be created. Close connections of the Slovene urban population with their rural home (with their roots) can turn the former rural-(sub)urban migration flow into a (sub)urban-rural one.

On the other hand, the growing importance of recreational activities in rural areas makes us aware of the need for the conservation and maintenance of cultural landscape. Slovene rural areas are potentially more interesting for the EU urban population than for the Slovene population. Almost all Slovene urban residents live in close vicinity of cultural landscape (the majority within 15 minutes walking distance) and have relatives in the countryside. For that reason farm (countryside), tourism has a very small (limited) market in Slovenia. The need for expansion of this market into the neighbouring EU countries with great urban agglomerations is therefore inevitable if we want to increase income in that field. Without proper marketing, possible only with certain co-ordination at the national level and with a strong support to the local initiative the natural beauty of the Slovene cultural landscape by itself represents a very small advantage.

Finally it should be said that the problems of the Slovene marginal rural areas are in many ways similar to most of the other European countries (Eastern and Western, Northern and Southern) but in some other ways they are also very specific. For that reason, we must not just transfer the developmental methodologies from the countries with a longer tradition in that field. These methodologies have to be adapted to our specific conditions and for the specific developmental problems, new authentic methodologies have to be designed. The present concept of integral rural development is only the basis (experience) for the formation of new and more effective concepts for a new rural development plan. The question of marginal rural areas is one of the crucial questions of further social economic and spatial development in Slovenia and has to be treated as such.

References

Drozg, Vlado (1999), *Opredelitev naselij upravičenih do statusa mesto* (The definition of the settlements that are justified to get the status of the city), Typescript of an expert valuation, Government of the Republic of Slovenia, The Service for Local Self-government, Ljubljana, p. 24.

Gams, Ivan (1991), The Republic of Slovenia - Geographical Constants of the New Central-European State, *GeoJournal* 24:4, pp. 331-340.

Haggett, Peter (1983), *Geography: a modern synthesis*, Harper & Row, New York, p. 644.

Wießner, Reinhard (1999), Ländliche Räume in Deutschland, *Geografische Rundschau* 51, H. 6, pp. 300-304.

PART 2
ENVIRONMENT AND SUSTAINABILITY

9   Establishing a sustainable conservation policy in northern Saudi Arabia

GARETH JONES AND AHMED AL MODAYAN

Introduction

Saudi Arabia has been transformed in the space of sixty years from a frag-mented, tribal society comprising migratory Bedouin tribes that survived on the products of oasis agriculture, to that of a cohesive, modern state with unimagined economic wealth based upon the seemingly limitless reserves of crude oil. The Bedouins adapted to the difficult arid conditions by mov-ing from oasis to oasis in an effort to optimize the availability of scarce natural resources. These people, numbering no more than 250,000, pos-sessed little in terms of material assets. Their well-being was dependent upon their religion and their ability to maintain their camels, flocks of goats and some sheep, as well as trading with the periodic caravans that linked the oases with the coastal areas. This way of life finally succumbed by the 1960s when the revenue from oil sales allowed a transformation of the so-cial structure for all Saudi inhabitants.

The physical environment

Considerably more than half of the area of Saudi Arabia is desert. The Rub Al Khali, the so-called Empty Quarter or Great Sandy Desert, extends over much of the south-east and beyond the southern frontier into the adjacent countries of Yemen and Oman, Figure 9.1. Still largely unexplored, the Empty Quarter has an approximate area of 777,000 km^2 (300,000 sq. miles). In the north of Saudi Arabia an extension of the Syrian Desert pro-jects into and extends south-east as an upland desert of red sand covering an area of about 56,980 km^2 (22,000 sq. miles), an area known as the An-Nafud. A narrow extension of this desert, the Ad Dahna, links An-Nafud to the Rub al-Khali. A central plateau region, the Najd, broken in the east by a series of uplifts, extends south from An-Nafud. A number of *wadis* (watercourses) dry except in the rainy season, traverse the plateau region. The western limits of the latter are delineated by a mountain range extending generally north-west and south-east along the eastern edge of Al Hijaz and Asir regions. The highest point in Saudi Arabia, Jabal Sawda 3,133 m (10,279 ft), is located in the south-west of the country. Between the western mountains, which have an average elevation of about 1,220 m (4,000 ft), and the Red Sea is a narrow coastal plain. In the east, along the Persian Gulf, is a low-lying region known as al-Hasa underlain by the major oil bearing deposits.

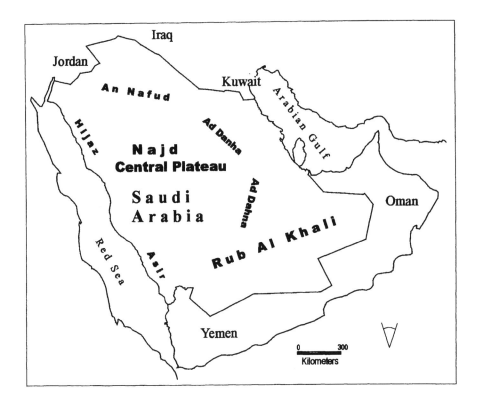

Figure 9.1   Main physical regions of Saudi Arabia

Climate

Extreme heat and aridity are characteristic of most of Saudi Arabia. The average temperatures for the months of January and July in Riyadh are 14.4° C (58° F) and 42° C (108° F) respectively. The average temperatures in Jeddah for the same months are 22.8° C (73° F) and 30.6° C (87° F). Average annual precipitation in Riyadh and Jeddah is 81 mm (3 in) and 61 mm (2 in), respectively. Evaporation negates the effectiveness of much of the meagre rainfall. Because of the general aridity, Saudi Arabia has few permanent water courses.

Natural resources

Prior to the discovery of crude oil in the Arabian Gulf the natural resource base was confined to a scattering of oases, most of which occurred north of Rub al-Khali. The larger oases were located above water-bearing rocks which allowed a natural upwelling of fresh, 'sweet' water to the surface. Simple technology allowed the construction of shallow wells (often no deeper than five metres), to gain access to the water. The combination of difficult geological conditions and the absence of suitable engineering technology prevented deeper wells from being constructed and consequently, irrigation was confined by the paucity of water. Elsewhere in Saudi Arabia, rain-fed agriculture occurred only in the extreme south west, in the Asir area. Sporadic rain-fed agriculture occurred along the sides of wadis and on the outwash plains where wadis spilled out of the mountains onto the lowlands. This was particularly the case along the western side of the country. Some larger tracts of semi-permanent pasturage occurred in Ad Dahna' and on the plateau region, see Figure 9.2.

Conservation policy in the Arabian Peninsula

Nature conservation has had a long tradition in Arabia. The Koran and Arabic poetic literature attach great importance to the value of preserving the natural heritage, embodied with the need for the traditional Islamic respect for nature (Ba Kader *et al.*, 1983).

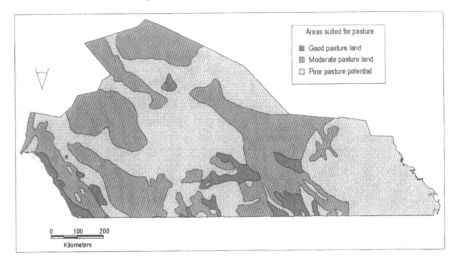

Figure 9.2   Areas of pasture in northern Saudi Arabia

Traditional forms of protected area or resource reserves known as *hema, hima, hujrah* or *ahmia*, have their origins over 2,000 years ago in the pre-Islamic period, and may have been developed as an acknowledgement of the scarcity of renewable resources in the difficult natural environment that prevailed throughout Arabia. The need for conservation and wise use of natural resources in order to support a sustainable rural economy became established as a fundamental element of Bedouin life style. In a major break with religious teaching the majority of *hemas* were gradually abandoned throughout the 1960s, following the opening of lands to free grazing by the decree of King Abdulaziz in 1953 (Abo Hassan, 1981; Abuzinada and Child, 1991; Draz, 1985).

In 1986 a Royal Decree led to the setting up of an 'adequate representative system of protected areas' (Child and Grainger, 1990) and the National Commission for Wildlife Conservation and Development (NCWCD) became responsible for the creation and maintenance of the protected areas system. Six categories of protection have been established:

- Special nature reserve;
- Nature reserve;
- Biological reserve;
- Resource use reserve;
- Controlled hunting reserve;
- Traditional *hemas.*

A total of 18 protected areas has so far been designated, only one of which has received the title of 'national park', the Asir National Park. The decade following the Royal Decree was occupied mainly by the completion of a series of specialized data collection initiatives and a programme of five year development plans has been established. At the beginning of the 1990s some 15 sites had been designated of which, four large areas were located in the northern area of the country. Table 9.1 provides details of these four regions and Figure 9.3 shows their location.

Table 9.1   Details of protected areas, northern Saudi Arabia

Protected area	Designation	Area (ha)	Date created
Harrat al-Harra	Managed Nature Reserve (IUCN Cat. IV)	1,377,500	1987
Tubayq	As above	1,220,000	1989
Khunfah	As above	2,045,000	1987
Sakakah	Non-hunting zone	10,780,000	?

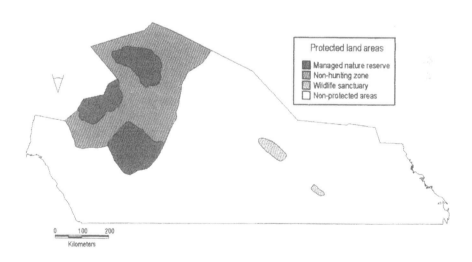

Figure 9.3   Protected land, northern Saudi Arabia

Although the location of the northern protected areas is well away from the industrialized and urbanized areas of Saudi Arabia, the designation of protected areas created a number of problems. Some of the problems can be recognized as universal problems found in most countries while others are specific to the special physical and social circumstances found in Saudi Arabia, see Table 9.2.

Table 9.2   Problems faced by protected areas in Saudi Arabia

Ubiquitous problems for conservation	Specific problems for conservation
• Population growth	• Replacement of migratory life style by permanent settlement
• Impact of industrialization	• Spread of oil-based industries and pipelines
• Problems of pollution	• Massive growth in motor vehicle pollution
• Growth of transportation	• Major road building programme,
	• Mass availability of all-terrain four-wheeled drive vehicles
• Intensification and spread of agriculture	• Development of centre pivot irrigation systems
	• Adverse climatic factors, drought
	• Shifting sands and large areas of bare rock.
	• Over-use of ground water table

The research conducted on this paper posed two basic questions:

1) How accurate had NCWCD been in identifying and establishing suitable areas of protected land;
2) How could GIS assist in identifying the suitability of designated protected areas through the availability of a comprehensive environmental data base.

Establishing a conservation data base

A considerable number of data sets relevant to the assessment of conservation potential and creation of protected land existed for Saudi Arabia. Unfortunately, the data sets have compiled by many different organizations and individuals and for many different purposes. Not unexpectedly, there was little similarity between the scale of mapping, the tolerance levels used when drawing category boundaries, or the map projection. Despite the inherent differences that existed between the various data sets, they represent a comprehensive collection of geographic data some of which might have potential for assisting with the location of conservation areas. Table 9.3 lists the main data sets and the origin of the data where known.

Table 9.3   Data layers used in the Saudi Arabian Environment Database

Description of data	Mapping potential
• Digitized outline of Saudi Arabia with international boundaries correct to 1999	• Basic map outline of Saudi Arabia
• Elevation. Contour lines supplemented by spot height elevation values	• Digital terrain model
• Digitized soil map. 59 categories of soils with 32 variables for each soil polygon	• Soil type map. Derived maps based on the 32 variables, e.g. land prone to flooding, land comprising rock outcrops at surface.
• Digitized vegetation map	• 17 categories
• Digitized agricultural land map	• 3 categories
• Digitized pasture land map	• 3 categories
• Digitized map of blown sands	• 3 categories
• Digitized map of wadis	
• Digitized map of conservation areas	• Three categories, some of which equated with IUCN conservation categories.
• Digitized map of aquifers and oases	
• Climatic data held in Excel spreadsheet. Detailed monthly statistics for 23 meteorological stations. Summary statistics for 40 other sites.	• Statistical data used to produced isoplethed maps of temperature, precipitation, air pressure.

Advantage has been taken of recent advances in GIS software and computer hardware to compile a data base of environmental variables relevant to the establishment of protected land areas and the preparation of a conservation strategy. This paper focuses attention on the area of Saudi Arabia north of latitude 26°. It is intended to extend the study to the whole of the country. Figure 9.4 shows the most detailed data set available, that of the soil data. The complexity of the original map required its compilation in full colour but due to publication restrictions has, of necessity, been reproduced here in black and white.

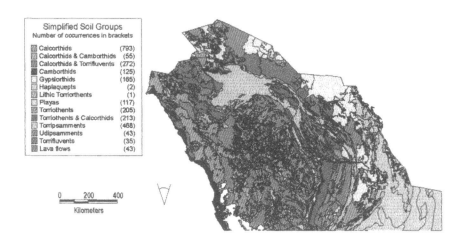

Figure 9.4   Soil types of northern Saudi Arabia

A data base comprising statistical and spatial data in electronic form permits subsequent flexibility in the analysis and interrogation of the data. GIS allows data sets to be overlain and customized maps produced which show specific combinations of data that has been selected from individual data layers. Buffer zones can be created around point features such as oases or settlements, corridors can be created along wadis and roads. Use of these simple techniques allows a new level of interpretation to be placed upon the designated areas of protected land and allows a new insight into the conservation value of the study areas.

To answer the first question posed earlier in this paper, in which the value of the existing protected land would be assessed, data was prepared in which protected areas, good agricultural land and good pasture land were overlain. In addition, settlement and roads and tracks with 10 km buffer zones were also included. The result of this overlay revealed that two very small areas of good agricultural potential fell within the established reserve areas. No major roads passed through the reserves and only one town, Tayma, was within 10 km of the edge of one reserve. Fifteen different soil types occurred within the established reserves and no relationship could be established between reserve areas and soil types. The conclusion can be reached that the location of the reserve areas were well located and did not conflict with settlement, routeways or land of economic value. Positioned in the northern territorial area of Saudi Arabia, the existing reserve areas and the no-hunting buffer zone were located in a sparsely populated area of Saudi Arabia and conflict between conservation and human population has been largely avoided. NCWCD can, therefore, be complimented on the location of the northern protected zones.

Turning to the second question, that of identifying additional areas for designation as protected areas, a series of criteria were established for the recognition of new sites. Based on contemporary IUCN policy, the aim was to identify new areas of conservation value that were as large and contiguous with existing designated areas and they should also be approximately spherical in shape. This would ensure a small as possible perimeter relative to the total area. A desirable feature would be for any new areas to act as corridors linking both established and new areas of conservation value. Any newly designated areas must be located away from settlements and major lines of communication.

Firstly, the data sets for soil types and existing areas of protected land were superimposed and a search made for the most typical soils within the areas of managed nature reserves. Soils in the categories Calcorthids and Torrifluvents (Category 3) and Torriothents and Calcorthids (Category 10) were most commonly found in two of the designated conservation zones. A search was made for contiguous areas of similar soil types and which also occurred adjacent to existing conservation zones. The resultant areas are shown in Figure 9.5.

Using this approach it was possible to show that two additional areas of land exist which could add substantially to land that could be designated Managed Nature Reserve Land (IUCN Category IV). The shape of the new areas did not approximate ideal shape of a sphere, and instead were broad linear areas that extended away from existing protected land. The area in the north east extended from the non-hunting area in a south east direct and shoed the beneficial feature of linking into the existing two small Wildlife Reserves. The new area extending to the south west of the region would extend from the exiting Managed Nature Reserve, and would approximately double its extent.

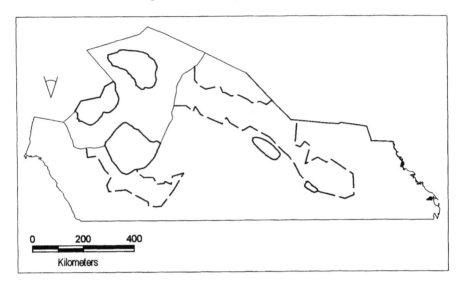

Figure 9.5   Possible extension of preserved land, shown by dashed lines

To assess whether the proposed new areas would conflict with existing settlement and roads, these features were overlain and a buffer zone of 10 km was designated to provide a zone of safety for the conserved areas. The resulting map is shown in Figure 9.6. Finally, a check was made to assess the likely conflict of the new areas with agricultural and pastoral land and the derived conflict map is shown in Figure 9.7.

The results shown in Figure 9.6 suggest that existing towns and routeways would not come into serious conflict with the proposed extensions for protected land. The final overlay introduced data for agricultural land and grazing land. Figure 9.7 shows that the potential conflict between land categorized as 'in agricultural use' was small, comprising three discreet units of land one of which was adjacent to a single area of good pasture land. It was this latter combination that was most serious as it prevented continuity between the proposed enlarged area of protected land and an existing Wildlife Reserve.

Two strategies become possible at this stage: local knowledge of the potential conflict would be required. It may be that the agricultural and pasture categories were of relevance in former times and that rural depopulation may have resulted in the underuse of both the areas. In this case,

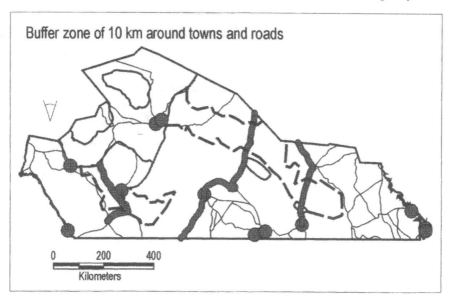

Figure 9.6   Overlay of roads and settlements to assess conflict with conservation zones

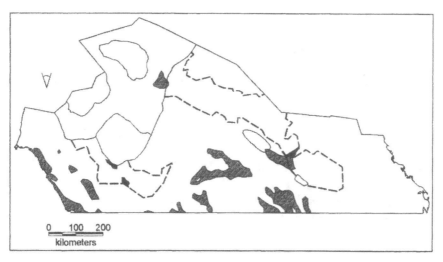

Figure 9.7   Conflict map showing overlap between proposed conservation areas (dashed lines), good agricultural land (black areas) and good pasture land (grey area)

conflict may not occur. Alternatively, a government subsidized 'set-aside' or 'land bank' policy could compensate the traditional users of the pasture and agricultural areas, or the local inhabitants retrained to work as conservation rangers instead of farmers. Figure 9.7 clearly shows the practicality of extending the existing areas of protected land. Not only would the area of protected land be doubled in extent, it would serve to link existing Wildlife Reserves by means of major corridors of protected land. The use of GIS has enabled the second of the objectives set out earlier in this paper to be positively answered.

Conclusion

The methodology described in this paper has shown that the conversion of analogue paper maps to digital format for subsequent use in a GIS can successfully identify potential areas of conservation land. In addition, the methodology can be used to verify decision making that has been made on traditional grounds when GIS was not available.

The method used to allocate land for conservation use has often been that of designating those areas that remain unused after agriculture, forestry, urban, industrial and transportation uses have taken their share of the land. Consequently, areas designated for conservation often occupy land with the least economic usefulness and it is often the case that conservation areas are assigned sub-optimum sites, those that are too rocky, too remote or otherwise unsuited for modern land use practice. There is increasingly a recognition that a protected land policy should be integrated into a national land use plan and should not be seen as an after-thought in land use planning.

Designation of protected land areas has assumed a distinctly proactive role in which land of high ecological diversity or of conservation value is now given equal priority to land which has been traditionally assessed as economically valuable to society. The increasing availability of skilled staff, trained for example as conservation and environmental planners, has enabled land to be retained for non-economic use to be assessed in a more rational manner. Modern technology in the form of Geographic Information Systems has allowed the re-working of much traditional information which has been translated from paper analogue format to digital format. The case study used in this paper has shown that Saudi Arabia possesses an advantage over many other nations in the opportunity for establishing new areas of protected land. Large areas of land are currently under utilized due to physical environment restrictions, most notably a lack of moisture. In addition, the rural population has declined rapidly as new employment opportunities have appeared in urban centres. Government attitude towards the establishment of new protected areas is also positive and as such progress in designating new areas of protected land is good.

The authors of this paper present the use of GIS as an additional tool in the identification of areas suitable for conservation. The authors do not recommend GIS as the only means of identifying areas suitable for conservation. Such an approach would be guilty of an oversimplification of a highly variable and complex situation. GIS can provide a method whereby different variables that exist in the form of data overlays, or coverages, can be interpreted in ways that would be difficult to achieve without the use of GIS methodology. This approach is recognized as supplementing, and not replacing, the traditional field work methods of the conservation biologist. In the context of Saudi Arabia, the methodology is especially useful due to the difficult physical conditions which make field work very difficult for many months of the year.

References

Abo Hassan, A.A. (1981), Rangeland management in Saudi Arabia, *Rangelands* 3(2), pp. 51-53.

Abuzinada, A.H. and Child, G. (1991), *Developing a system of protected areas in Saudi Arabia*, National Commission for Wildlife Conservation and Development, Riyadh, Paper presented at the Third Man and Biosphere Meeting on Mediterranean Biosphere Reserves and the First IUCN-CNPPA meeting for the Middle East and North Africa, 14-19 October 1991, Tunis.

Ba Kader, A.B.A., Al Sabbagh, A.L.T.S., Al Glenid, M.S. and Izzidien, M.Y.S. (1983), 'Islamic principles for the conservation of the natural environment', *IUCN Environmental Policy and Law Paper* No. 20.

Child, G. and Grainger J. (1990), *A system plan for protected areas for wildlife conservation and sustainable rural development in Saudi Arabia*, NCWCD, Riyadh.

Draz, O. (1985), 'The hema system of range reserves in the Arabian Peninsula, its possibilities in range improvement and conservation projects in the Near East', in J.A. McNeely and D. Pitt (eds.), *Culture and conservation: the human dimension in environmental planning*.

World Conservation Monitoring Centre (1999), *Protected Areas of the World, A Review of national systems in Saudi Arabia*, http://www.wcmc.org.uk/cgi-bin/pa_paisquery.p.

Acknowledgements

Grateful thanks are expressed to both the University of Strathclyde, Glasgow and the University of King Abdulaziz, Jeddah for financial support which allowed the compilation of the data base necessary for this research project. Sayeed Ahmed and George Yule have provided technical assistance with digitising and GIS software.

10 Biosphere reserves – sustainable development of marginal regions?

WALTER LEIMGRUBER AND THOMAS HAMMER

Introduction

Projects on sustainable development are first and foremost plans, ideas, hopes, although they profit from the current concern about the state of our environment. It is true that transformations of the environment by natural and human processes have occurred throughout history, but the 20th century has been characterized by an increasingly violent impact of human actions:

> Nowadays, human activities are altering these flows [of energy and material; WL] at unprecedented scales; human-induced consumption and transformation of net primary productivity is estimated to be about 40 per cent of that carried out by the Earth's terrestrial ecosystems (Stanners and Bourdeau, 1995, p. 9).

It may therefore make sense to develop ideas about how to protect at least certain regions of the earth from total degeneration - after all, humanity is increasing in number, so we have to live *with* the environment.

Environment as a global issue

As an ecosystem, the Earth is unique within the solar system. The conditions of the atmosphere are such that the mean temperatures are sufficient for life. Everyone is aware of the delicate balance reigning between the various factors composing our environment, as we are also conscious of the particular role humans have been playing. When they started to define their needs according to subjective wishes (which went beyond physiological requirements), their impact on land, water and air has increased at an alarming speed. As a consequence, it has become evident that human influence is present around the globe: no place is devoid of a minimum human impact, natural landscapes are something of the past; cultural landscapes have taken their place.

Everybody is concerned by the state of the environment, every individual as well as the global society (the international community) are called upon to act: all scales are touched, albeit to a different degree. The globalization of environmental issues is amply demonstrated by the Ozone depletion in the stratosphere above Antarctica, and the transportation of air- and waterborne pollution across oceans and continents, but we are also confronted with regional issues such as the pollution of the North Sea and the Mediterranean, or with local ones such as the illegal dumping of toxic waste.

Transformations of the cultural landscape

Landscape, a term central to geography, has both an objective spatial context and subjective meanings. Grandgirard and Schaller (1995, p. 30) call it 'a portion of space, perceived by a particular observer at some specific moment from a certain viewpoint on the surface of the Earth' thus defining it as a partial and subjective representation of space, viewed from physiological, sociocultural and personal perspectives. Landscape is the result of human perception and action within a given spatial context. The result of human interference can be summed in the (value-free) term *cultural landscape*. The first cultural landscape, when human societies started to transform nature intentionally, was obviously very close to nature, but throughout human history, the impact of modern technology (from the ox-drawn plough to the Concorde) has led to a growing artificialization, and the cultural landscape close to nature has gradually turned into civilized and technical landscape (Figure 10.1).

Landscape transformation by man is a process guided by human perception and values. More precisely, it is the different actors (i.e. individuals, groups, communities, societies) according to their social, economic and political affiliations who take decisions based on their respective value systems, oscillating between secular and sacred attitudes (Leimgruber, 1998, p. 30). Townscapes, industrial landscapes, farming landscapes etc. reflect such decisions, depending on judgements about a region's potential and suitability for its particular utilization.

During the last generation (c. thirty years), the human society in the 'North' has initiated reflections on the consequences of the processes

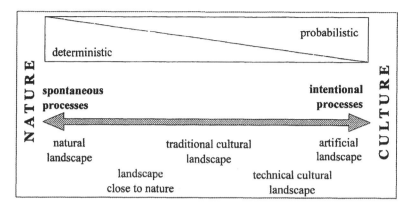

Figure 10.1   The landscape continuum

which have been going on since the industrial revolution, in particular about the increasing intensity of its impact on landscapes. The loss of biodiversity, the degradation of arable land, the growth of pollution and of the garbage mountains, the awareness of the limits to resources (e.g. water) and economic growth have led to a new attitude towards the environment. In recent years, ideas such as a carbon-dioxide tax and the 'polluter-payer' principle have ceased to be a taboo in political discourse. True, we are still far from concrete measures, but if it is to take another generation until they are implemented and become effective, it may really be too late.

About the same time, another awareness has manifested itself. The gradual loss of cultural diversity as a consequence of increased mobility, modern information technology, and globalization have made us aware of the fact that we are losing our roots. A new form of cultural imperialism is emerging, epitomised by terms such as 'macdonaldization', 'pizza-globe' or 'coke-world'.

The UNESCO biosphere reserve programme

Following the 1968 UNESCO Conference on the 'Conservation and Rational Use of the Biosphere', the Man and Biosphere (MAB) Programme was launched in 1970 on a world-wide scale. Out of it grew the biosphere reserve concept, formulated in 1971 and subsequently developed and put into action (Table 10.1). Its objective was to strike a balance between conserving biodiversity, encouraging economic and social development, and preserving cultural values and variety. A biosphere reserve is based on the interaction of the natural and the cultural systems; it admits that nature protection alone cannot be an aim in a world whose population is growing at an accelerated pace. Besides, humans are recognized as being part of the biosphere: not only as social beings but also as part of nature - a fact, former generations had forgotten. This is the real originality of the concept: the fact that it attempts to marry nature with human activity, i.e. to combine the protection of and the respect for the natural ecosystem with man's attempts to make a living.

Table 10.1   The evolution of the biosphere reserves idea

1968   UNESCO Conference on the 'Conservation and Rational Use of the Biosphere

1970   Foundations of the UNESCO research programme «Man and Biosphere» (MAB)

1971   Biosphere Reserves created as a MAB-policy objective

1972   United Nations Conference on the Human Environment, Stockholm

1974   Development of the concept of Biosphere Reserves

1976   Initiation of the Biosphere Reserve global network

1983   Minsk conference (UNESCO & UNEP) on Biosphere Reserves

1984   Action plan for Biosphere Reserves

1995   Sevilla strategy: reformulation of goals, consistent criteria established

In October 1999, UNESCO listed 357 reserves in 90 countries on all continents (List ..., 1999), and their number is growing. Many of these reserves are situated in what we would call marginal regions, and they often originated in national parks and nature reserves.

Biosphere reserves are constructed on a three-zone basis (Figure 10.2). In this way, a sudden break between an area of total protection and surrounding areas of intensive land use can be avoided. This structure is more than a simple convenience for management purposes (as could be guessed from the description furnished by Conservation·International (CI), 1998a): it enables the full integration of a biosphere reserve into the cultural landscape. The three zones, in particular the core, need not be contiguous. By offering this kind of flexibility, areas meriting total protection from human interference can be determined according to local circumstances; while the buffer still guarantees for limited and controlled human impacts.

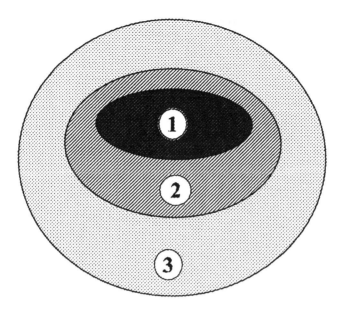

Figure 10.2 Model of the UNESCO biosphere reserve
1) core zone; 2) buffer zone; 3) transition zone

The three zones can be defined as follows (CI, 1998a):

1) *Core zone*: strictly protected areas with little to no human influence. They are used to monitor natural changes in representative ecosystems and serve as conservation areas for biodiversity;
2) *Buffer zone*: areas surrounding the core zone where only low impact activities are allowed (research, environmental education, recreation);
3) *Transition zone*: the outer zone where sustainable use of resources by local communities is encouraged.

While the core is strongly protected, settlements and a number of human activities are tolerated in the buffer and transition zones (e.g. environmental training, ecological observation, soft tourism, extensive traditional and/or intensive sustainable farming). Intensive cultivation is located outside the biosphere reserve. Economic development is based on the idea of sustainability, adopting a long-term vision towards an equilibrium. Ecologically motivated research and ecological observation should guarantee the survival of the biosphere reserve into the future.

Each biosphere reserve has three basic functions (UNESCO, 1996, p. 4; CI 1998 a):

1) *A conservation function*: to contribute to the conservation of landscapes, ecosystems, species and genetic variation;
2) *A development function*: to foster economic and human development which is socially and ecologically sustainable;
3) *A logistic function*: to provide support for research, monitoring, education and information exchange related to local, national and global issues of conservation and development.

These functions were defined during the discussions, which led to the Sevilla strategy (1995), a document favouring collaboration among local communities, government institutions, NGOs and private enterprises. By emphasizing co-operation between public and private partners, the responsibility for our ecosystem has been allocated to all (including the economic) actors. This is a very important point, which has recently received support by the UN General Secretary, Kofi Annan. Addressing the managers of the world during the World Management Forum in Davos in January 1999, he said: "I call on you, individually through your firms, and collectively through your business associations, to embrace support and inact a set of core values in the areas of human rights, living standards and environmental practices" (Gschwend *et al.*, 1999, p. 1/1). Quite clearly, he made the link between social and environmental problems.

The biosphere reserve concept as such looks very promising indeed, and it is likely to improve the integration of cultural and natural landscape, of human activity and the protection of the environment. The pendulum may swing back from essentially intentional towards spontaneous processes, resulting in a re-naturalization of the landscape.

The Entlebuch project

Location and characteristics of the area

Switzerland is not a newcomer to the biosphere reserve idea. In the 1970s already, the Swiss National Park in the Grisons was registered, and it fulfilled the requirements of that time. However, since the Sevilla strategy has been implemented, it has been cancelled from the list as it no longer respects the more severe criteria. A new project in an extended region around the National Park is currently being developed.

The study area of this paper (Entlebuch valley and district in the canton of Lucerne; Figure 10.3) is a microcosm in central Switzerland. It straddles the limit between Alps and Plateau and is characterized by a very

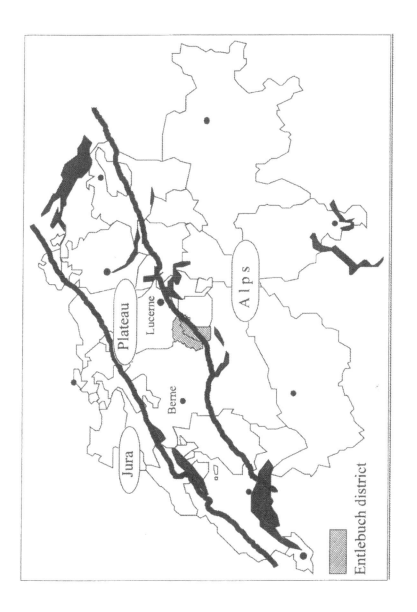

Figure 10.3 **The Entlebuch region in the Swiss context**

lively topography. With a surface of 410 km², it measures 1% of the whole country and 27% of the Canton of Lucerne. The (catholic) Entlebuch district borders on the (protestant) canton of Berne, it is a border region characterized by cultural differences.

As many other mountain areas in Switzerland and elsewhere, the Entlebuch has experienced a considerable out-migration throughout this century, in particular after World War II. Between 1950 and 1990, the district lost 5.5% of its population (Table 10.2), and only three out of 9 communes saw the number of their inhabitants grow.

Table 10.2   Population growth in the Entlebuch district, 1950-1990

Region	1950	1970	1990	50-90%
Entlebuch district	18450	18001	17431	- 5.5
Canton of Lucerne	223249	289641	326286	+ 46.2
Switzerland	4714992	6269783	6873698	+ 45.8

Source: Swiss Population Censuses

Out-migration in mountain regions has a long tradition, and the Entlebuch district is no exception to this. It has promoted the process of ageing, although a relatively high fertility has compensated it to some extent. Since 1950, the percentage of people above 65 years of age has doubled and reached national average, whereas the proportion of the young population has diminished by one third, but still remains well above the Swiss mean (Figure 10.4). It is the young generation, which tends to depart, and one can say that the Entlebuch has been furnishing manpower to external labour markets (especially the Lucerne conurbation) for quite a time. The district itself offers far too few job opportunities for the young generations.

The economic potential of this rural area is limited. Extensive farming (dairying and cattle breeding) and forestry dominate, the few industries are mainly resource-based (wood-processing). Tourism has become a supplementary source of income since the early 1950s, but its role is limited. Winter tourism operates under conditions of uncertainty given the relatively low elevation (between 800 and 2000 m above sea level): similar to other stations in the Pre-Alps, its long-term future depends on the degree of global warming. Thus, there are few economic perspectives for the local population, but on the other hand, the environmental quality is high. The region may be economically marginal but it can be called ecologically central (Leimgruber, 1994, p. 9).

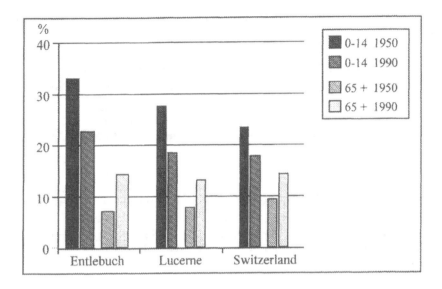

Figure 10.4   Young and old age groups in %, 1950 and 1990

Source: Swiss Population Censuses

The specific natural features

As a whole, the Entlebuch landscape can be classed as being very close to nature, although at a larger scale, intentional elements dominate to a varying extent, giving rise to technical cultural and sometimes even to artificial landscapes. Among the many natural features, two merit to be highlighted: forests and wetlands. The Entlebuch region is characterized by a high percentage of *woodland* (42%, in 1997, against a 30% Swiss average) a figure, which has risen considerably since 1952 (29% - Swiss mean 23%). This increase is due to both afforestation projects and natural reforestation. The former had to replace the woods cut for fuel used by glass industry, lactose production, and charcoal burning in the 18th and 19th century, the latter reflects the decline of agriculture: abandoned marginal lands return to woodland within a few years. The whole region benefits from the increase in forest area: not only is there a considerable resource base for the local wood industry, but the forests can go on to play their ecological role of stabilizing slopes and regulating drainage. Occasional avalanches close to skiing areas provide future afforestation projects with an additional incentive.

Another special feature are the numerous *bogs,* occurring in an area with between 1400 and over 2000 mm annual precipitation. Due to coal shortage, many of the lesser bogs have been exploited during World War II, others have been drained for the purpose of farming, but still a fair number has survived and is now safe from degradation. Since 1987, an amendment to the Constitution offers protection on a national scale.

The cultural landscape: heritage and current situation

From the point of view of its cultural background, the Entlebuch has no spectacular historic heritage to offer. Churches and chapels are good examples of rural religious art and architecture, and traditional farmhouse architecture still persists. The aspects of the villages, of course, have changed with the transformation of activities, and in some cases, the results are far from satisfactory.

Apart from a number of architectural atrocities in the main valley and of the tourist development in Sörenberg (a lateral valley), spontaneous traits still dominate: roads and railway follow natural guidelines, most settlements stick to their natural and usually well protected sites (with the exception of a second home zone in Sörenberg, built right below a major landslide), and the extensive way of farming does not harm landscape too much. Recent legislation favours environmentally inoffensive land-use practices, thereby adding to the relatively natural outlook of most of the Entlebuch district.

In 1999, the cultural landscape of the Entlebuch appears to be in a better condition than one hundred years ago when the destruction of the forests was well visible and when the downstream areas almost as far as Lucerne suffered from frequent inundations due to deforestation.

Regional efforts towards the creation of a biosphere reserve

Three questions can be asked in the context of this example:

1) Can this valley be developed within the biosphere reserve concept or is it either too small or too specific to be included in this world-wide strategy?
2) Is it representative of Switzerland?
3) Can it serve as a model for other projects in Switzerland, in Europe, or in other continents?

Before attempting to answer these questions, we shall discuss a few additional issues concerning the project.

The creation of a biosphere reserve is above all a social and political challenge, concerning different institutional levels (Leimgruber and Imhof 1998, p. 391). This is particularly true in the case of Switzerland where all projects have to obtain consensus among the people. A final political decision will have to be voted. Due to the many laws and ordonances, the preparation of such a project requires a lot of time and patience; many people need to be convinced that sustainable development and environmental projects are useful (even if no monetary value can be quoted) and in their own interest, but there will always remain a small group of notorious opponents to such a venture.

In this respect, the Entlebuch biosphere reserve project is not different from projects like nature protection, and the regional management group working on it faces no easy task. However, it can build on solid ground: the protection of bogs since 1987 and the regional development concepts by the LIM-region concerned (see Leimgruber 1985) had prepared the field. In 1996, a regional marketing organization (REGIOPUR) was founded, aiming at promoting regional products. In the following year, the regional planning group and the Swiss Foundation for Landscape Conservation initiated a joint project, «Entlebuch life-space», which anticipated the biosphere reserve idea. The management group set up for this purpose was given the task to tailor the biosphere reserve concept to the specific case of Entlebuch within three years (1998 - 2000), a short time to prepare the project and convince the population; after all, a change of mentality cannot be imposed from above but has to be reached by slow and painstaking discussions.

The project requires a huge effort to convince the regional population because a biosphere reserve cannot be decreed from somewhere above but has to be supported by the people concerned. The population has to understand the project's true significance: the Entlebuch will not be transformed into a 'reservation' where tourists arrive to take pictures of the last 'natives' still alive, but it is to remain an active region with the particularity of long-term sustainability covering both natural and sociocultural elements. Human interventions will have to take this new (in reality old) philosophy into account. Paradoxically, the problem lies with nature and landscape protection: federal and cantonal laws protect about 50% of the whole area, and danger zones cannot be used for economic activity - people already feel overprotected. Hence, the creation of a biosphere reserve requires above all a change in thinking, in outlook, in mentality, in values.

The promoters in the regional management group have understood this problem. They spend hours explaining the project to people during communal assemblies, in meetings with specific actors (local and regional politicians, farmers), and through media campaigns. Their principal topics are:

1) Landscape is a marketing argument for tourism
2) Biosphere reserves are an integration of natural and cultural elements (animals, plants, humans with their needs)
3) The Entlebuch still excels by its biodiversity
4) People should find ways to use the land without contributing to its degradation
5) The notion of 'reserve' signifies economic and cultural development and modernization without the destruction of the natural bases of life.

Such arguments sound very theoretic and may not convince everybody. Usually, voters want to know the reality, e.g. the surfaces, which will be excluded from economic activities. Fortunately, this problem can be solved according to circumstances, as there are no specific UNESCO requirements for the surfaces of the individual zones. In the case of the Entlebuch, about half the area already enjoys legal protection of one sort or another, hence there will be no additional source of conflict about the designation of the core area. Inevitably, however, compromises will have to be found.

We can safely assume that most of the Entebuch biosphere reserve can be used much as it has been until today. For future development, tourism has been designated as leading factor, serve as the focus for a regional network which includes farming and the marketing of regional farm products. The old label (REGIOPUR) will be replaced by a new one ('Entlebuch Biosphere Reserve') in order to promote the concept within the whole of central Switzerland.

Conclusion

We can now try to answer at least partially the three questions formulated above.

1) As to the problem of size, let's remember that UNESCO never fixed any surface criteria for a biosphere reserve. They range from about 10,000 to over 1 million hectares (CI 1998 b). In Germany, the standard area is about 30,000 hectares, whereas Switzerland aims at 25,000 ha, of which 5% are to be allocated to the core zone (Germany: 3%). The Entlebuch district

measures 41,000 ha, i.e. only part of it may have to be included in the project.

The concept as such is sufficiently neutral to allow for small areas as well as very large ones to be classified as reserves, provided they apply the criteria correctly. National legislation has to safeguard that the separation between protection and unhampered human activities is made in a way, which does not harm the core zone and reduces the impacts in the buffer zone to a minimum. Since notions such as 'to harm' or 'minimum' are normative, the true impact may only be known after a long period.

2) As to the question of representativity, the Entlebuch stands for one particular landscape type of Switzerland: the Pre-Alps, used by man in a rather extensive way. Due to its economic marginality, it also represents the less well off regions in Switzerland.

3) The model function is emphasized by the regional management group. Nature protection and economic development, in particular (soft) tourism and marketing of 'ecologically' produced food are the focus of this function.

Indeed, the Entlebuch biosphere reserve could serve as a model on different levels:

1) Regionally: to rally the population around one common project, which could prevent further outmigration and increase the chances of survival of the population in its familiar surroundings. Tourism and regional products can be one element.

2) Nationally: to promote 'alternative thinking', i.e. an integrative mentality which sees humans and nature as inextricably linked, where sustainable development is not simply a keyword but can be lived. The biosphere reserve label for both the tourism region and the agricultural products could act as catalysts.

3) Internationally: to promote the bottom-up philosophy by demonstrating how such a project can be brought home to local populations who, in their mind, are far away from international organizations and their thinking and language, but who can learn through good and persistent information. The need to anchor such projects with the people concerned and to ask (and accept) their advice could be diffused in this way.

The idea of biosphere reserves may first of all appeal to urban populations, which have lost contact to nature. However, UNESCO's objective is not only to serve the urbane people but also to assist the rural people and at the same time ensure biodiversity and long-term survival of regional ecosystems (UNESCO, 1996). This requires an integrative way of thinking and a new attitude towards our environment: it must no longer be considered a 'quarry' where everybody fetches what they need but it is the common heritage of mankind which has to be passed on to future generations (if our generation wants them to survive!).

Further ideas for biosphere reserves have surfaced in Switzerland (Figure 10.5). Most of the regions proposed are very attractive and represent distinct landscape types, but we cannot transform Switzerland into one huge biosphere reserve. Besides, the road from the proposal to a reserve is long and full of obstacles.

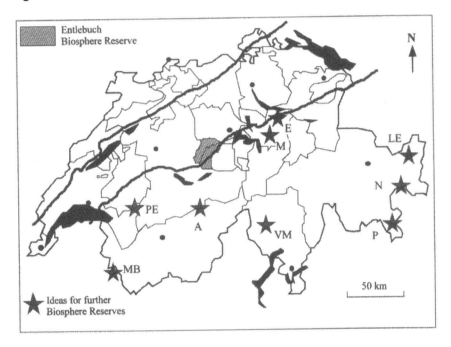

Figure 10.5   Location of potential biosphere reserves in Switzerland (1999)

MB: Montblanc region, PE: Pays d'Enhaut, A: Aletsch area, VM: Vallemaggia, M: Muothatal, E: Einsiedeln, LE: Lower Engadine, N: the enlarged National Park, P: Valle Poschiavo

Biosphere reserves: the ultimate remedy to save marginal areas from disappearing entirely from the map? Yes and no. If a region has become (economically) marginal, the creation of a biosphere reserve may be a blessing. However, too much success may turn the remedy into a curse - it is difficult to keep the balance between its potential attractivity and the goal of sustainability. Propagating an area on a large scale may lead to an increasing influx of visitors, as a result of which the equilibrium sought for will eventually be disrupted. Looking at it positively, under current circumstances we are optimistic for the Entlebuch project. The pendulum is about to swing back from secular to sacred values.

Biosphere reserves are the result of a change in attitude, in perception. This is an important point, which we geographers have to bear in mind. Therefore, let us conclude with a reference to an eminent geographer who has almost been forgotten, but whose fundamental work has recently been translated into English. In his *Pure Geography,* the Finnish geographer Johannes Gabriel Granö (1882-1956) emphasized the importance of perception in geography in the first paragraph of his book:

> The notions that we possess of places and regions gained by personal observation are derived from the overall picture provided by all our senses, the validity of which depends on both the extent of the area we have perceived through our senses at any one time and the duration and degree of detail of this observation. The total impression obtained has a definite extent in both space and time, in that our faculty of sight determines its size and the duration of our lives its temporal boundaries (Granö, 1997, p. 9).

The validity of the biosphere reserve concept and its importance for marginal regions must also be judged from this phrase.

References

CI (Conservation International Foundation, Washington, DC; 1998a), *The Biosphere Reserve Concept,* http://www.conservation.org/science/cptc/capbuild/unesco/what_is.htm.

CI (Conservation International Foundation, Washington, DC; 1998b), *Participating Reserves,* http://www.conservation.org/science/cptc/capbuild/unesco/res_tbl.htm.

Granö, J.G. (1997), *Pure geography,* ed. by O. Granö and A. Paasi, translated by M. Hicks, Baltimore and London, The Johns Hopkins University Press.

Gschwend, H., Mugglin, M. and Müller, S. (1999), *Wirtschaft und Ethik,* Doppelpunkt SR DRS 1, 7.3./14.3./21.3./28.3.1999, Manuskript.

Leimgruber, W. (1985), 'What is a mountain region?', *Innsbrucker Geographische Studien* 13, pp. 99-107.

Leimgruber, W. (1994), 'Marginality and marginal regions, problems of definition', in: Chang-Yi D. Chang (ed.), *Marginality and development issues in marginal regions,* Proceedings of the IGU Study Group 'Development issues in marginal regions', National Taiwan University, Taipei, pp. 1-18.

Leimgruber, W. and Imhof, G. (1998), 'Remote alpine valleys and the problem of sustainability', in: L. Andersson and T. Blom (eds.), *Sustainability and development. On the future of small society in a dynamic economy*, University of Karlstad, Karlstad, Sweden, pp. 385-396.

List of the world network of Biosphere Reserves (October 1999), Internet source at, http://www.unesco.org/mab/brlist.htm.

Schaller, I. and Grandgirard, V. (1995), 'Espace et paysage, deux concepts-clé de l'approche géographique', *UKPIK, Cahiers de l'Institut de Géographie, Unversité de Fribourg, Suisse*, vol. 10, pp. 25-37.

Stanners, D. and Bourdeau, P. (1995), *Europe's Environment, The Dobris Assessment*, Copenhagen, European Environment Agency.

UNESCO (1996), *Réserves de biosphère, la stratégie de Séville et le cadre statutaire du réseau mondial*, UNESCO, Paris.

Acknowledgements

We should like to thank the Entlebuch Regional Management (especially Dr. E. Ruoss and Mr. T. Schnider) for lively discussions and information material, Mrs. Iris Heller-Kellenberger (Zürich-Birmensdorf) for her helpful comments, and the 56 students participating in the Entlebuch field study (May 1999) for their enthusiasm.

11 Sorghum based farming systems in Botswana – the challenges of improving rural livelihood in a drought prone environment

CHARLES E. BUSSING AND DAVID W. NORMAN

Background and paper objectives

Any cursory look at the biophysical and socioeconomic data for Sub-Saharan Africa (SSA) as a whole immediately gives rise to two fundamental conclusions. The first is the degree of heterogeneity that exists in the region in terms of resource endowment, institutional and infrastructure development, political stability and maturity, and commitment to development. The second is that in terms of almost any economic developmental indicator, SSA as a whole, is continuing to deteriorate relative to any other region of the world (IITA, 1994, p.48; CGIAR, 1994). Given the diversity in SSA there are promising signs of progress in agricultural and economic development in a few countries, but in general the immediate and long-term prospects are not promising. Even more depressing is the fact that about 1.5 billion hectares or 80% of Sub-Saharan African dry lands and rangelands show significant signs of desertification, while 34% of the remaining land in Africa is reputedly under threat of desertification (Rwegayura, 1992; Agence France Presse, 1993).

Together with people in South Asia, people in SSA remain the poorest in the world. Unfortunately, in contrast to progress in South Asia in reducing poverty, the proportion living in poverty in SSA continues to increase. In fact the number living in poverty increased almost 1.5% between 1985-1992 (World Bank, 1995). Although in recent years there have been modest increases in gross domestic product (GDP), partly as a result of structural adjustment policies being implemented by many countries, the

145

result in per capita terms has been negative (i.e., because of continuing high population growth rates).

The continuing inability of the non-agricultural sector to provide gainful employment for the rapidly rising population of SSA means the proportion of the population living in rural areas is still very high. The success of the Green Revolution in many parts of Asia contrasts with the much poorer performance of agriculture in SSA. All indicators of improved food production per capita in SSA continue to show a declining trend.

Thus issues relating to both food security and sustainability are becoming increasingly important. Assuming human fertility levels remaining at projected levels, some estimates (IITA, 1994, p.5) indicate that by the year 2020 the food gap (i.e., the difference between food requirement and food production) in Africa will amount to an annual deficit of 243 million tons if food production increases at 2% per year. If food production increases at 4% per year an annual surplus of 49 million tons would be produced. However, based on past performance, increases at the latter rate are unlikely. Thus problems of attaining food security will become increasingly serious unless population growth is reduced substantially and the non-agricultural sector develops rapidly and generates significant foreign exchange earnings to permit substantial imports of food – an unlikely development.

The poor developmental record of SSA, as the above discussion has implied, obviously has been shaped by many factors, both human and natural in origin. For example, the natural resource base in much of Africa, is relatively unfavorable. In particular, with reference to the specific focus of this paper, much of Africa is in the semiarid ecological zone, which tends to be particularly prone to drought. The relatively unfavorable situation in Africa is illustrated by the fact that, for low-income countries as a whole, 18% of the land area is located in the semiarid ecological zone, while in Africa 52% of the area is in that zone (FAO, 1987).

The objective of this paper is, using Botswana as an example, to address selected issues related to rural development in a marginal environment. The first part of the discussion provides a context and describes the traditional sorghum based farming system that used to exist and which generally was well adapted to the harsh biophysical environment of the tropical savannahs. The second section suggests some strategies that will be necessary if rural livelihoods are to be improved in a sustainable manner.

National context

Botswana is distinguished in the context of Southern African political geography for its stable political history and steady economic growth since independence. The country is sparsely populated relative to its size (an estimated 1999 population of 1.6 million persons within an area of 561,800 sq. km.), but its semiarid, drought prone climate and limited surface and underground water resources present severe constraints to growth. Despite the rapid growth of the mineral sector since independence, livestock production and agriculture generally will continue to be the mainstay of the rural economy, particularly in terms of personal income and employment.

Approximately 71% of Botswana's land area is subject to traditional or 'tribal' rules of tenure. District Land Boards hold land in public trust for district-based tribal groupings. As in many other systems of African land tenure, tribesmen are assured by birthright access to sufficient free land to meet homestead and subsistence agricultural requirements. Livestock are typically grazed on communal or open lands, though private ownership of water rights often determines usufructuary grazing rights. State ownership extends over another 23% of Botswana's land area, consisting mostly of wildlife parks and reserves and leasehold ranches. Freehold tenure, mostly involving commercial ranching operations, claims another 6% of the nation's land.

The relationship between the environment and development in Botswana has attracted a lot of attention for many years. Three major factors have stimulated this interest. First, the economy of Botswana has a very high dependence on natural resources. Mining alone accounted for 45.1% of the Gross Domestic Product (GDP) in 1987/88, and has contributed between 80 and 90% of the value for the country's exports for many years. Although agriculture was only 3.5% of GDP in 1987/88, it is the major means of livelihood for over 80% of the country's population. Secondly, in recent years there has been increased pressure on the country's environmental resources with production from the land. Botswana's brittle environment has recently shown signs of strain from these demands. Thirdly, it has been realized that the future of a sustainable development effort will depend more on renewable resources and less on mining.

Given the current level of development in Botswana, the land provides the most important of all renewable resources alternatives for development. Increased production to raise the contribution of the land and employ more people is required if this resource is to provide any significant solution.

In previous decades, wherever increased production was required, the government of the day simply encouraged farmers to bring formerly unused land into productive use. In the 1950s for example, with a series of good rains following a drought, local farmers greatly increased their output and reduced the country's import of maize through clearing and planting new areas. New lands are no longer available in wetter regions of the country. Similar encouragement with heavy subsidies in the 1970s and 1980s has extended agriculture into marginal areas where questionable practices are being used. At the same time there has been environmental deterioration mainly due to increased populations of both people and livestock on the land communally owned and grazed for many decades. In the last decade and a half researchers and policy makers have carefully rethought issues involving growth, development, and environment.

The traditional farming system

Botswana is a country that offers far from ideal conditions for arable agriculture. The most severe farming problems generally stem from limitations imposed by a lack of water. The seasonality of rainfall and its low annual totals impose serious constraints on growing crops. In addition to low annual totals (typically 400 600 mm) there is a large inter and intra-seasonal variability of rainfall. This means that the farmer's ability to undertake timely operations for ploughing, planting, weeding, etc., in order to make more water available for plant growth and to improve the efficiency of water use becomes a key management variable. Rainfall in no month exceeds potential evapotranspiration and there is no way to predict rainfall after planting which is an important determinant of yield. The intra-seasonal variability of rainfall means that management is important; however, luck often plays a disproportionate role in the production equation.

Botswana farmers face two significant temperature problems. The months when most planting is concentrated because there usually is the greatest rainfall are also the hottest months. Soil temperatures often exceed the cardinal level for sorghum seedling viability. A second problem is the onset of cool nights at the end of the season. If there are even short drought periods, during which sorghum stops growing, late January and February plantings may not have enough time to mature. Soils are generally of poor quality, suffering from phosphorus deficiencies and low levels of organic matter. Also, soils tend to have a low water holding capacity and are subject to crusting following heavy rains. Soil erosion is a major problem particularly where many fields have been continuously cultivated for several decades.

The natural environment therefore makes crop production a high-risk business, and this fact is reflected in the farmers' outlook on crop farming, which is one of low input. The low input strategy is understandable when one considers there is a chance the farmer might lose the whole crop. Thus for the majority of farmers in the communal lands areas, the main objective in their crop enterprise is to simply produce enough food to feed the household.

Farming in a typical household is carried out in three separate locations. In the traditional Tswana pattern people belong to a central village. With the onset of the rains about October part of the family moves out to the lands to plough and plant and remains there until the harvest is in, while others reside at a cattle post to tend their animals. Most of the land area is subject to tribal rules of tenure, and tribesmen are assured by birthright access to sufficient land to meet homestead and subsistence agricultural requirements. Livestock are grazed on communal or open land.

Farms usually range from 1020 hectares, not including communal grazing areas, and consist of one or two contiguous blocks of land with 5 to 6 hectares under cultivation at any one time. The usual pattern of planting is to hand broadcast a mixture of seeds onto unploughed land and then turn the seed under by ploughing with a single furrow, ox-drawn mouldboard plough. Post-planting operations generally are confined to one weeding and bird scaring.

Few crops are, or can be, grown in the harsh climatic environment of Botswana. The principal crops are sorghum, maize, millet, beans, and cowpeas. Yields per hectare are extremely low averaging about 200 to 300 kilograms per hectare. Sorghum serves as the main food and cash crop. Relatively few farmers grow special cash crops just for sale.

A significant division of labour exists. Adult females provide most of the labour for sorghum production, as well as for household maintenance activities. Males traditionally have been responsible for tending cattle. Animal traction plowing is the only arable farming activity dominated by males. Because two-thirds of the active rural population attend school, tend livestock, or are engaged in off-farm wage employment, the relatively large amounts of land cultivated per family member imply the use of a labour saving farming systems management strategy.

Farming strategies can be significantly affected by the ability of farmers to withstand production shortfalls. Botswana farmers have a much greater ability to deal with crop failures because of their control of cattle assets. While many Botswana farmers do not own cattle, more than half are able to draw against their cattle inventories in the face of adverse circumstances, thereby stabilizing their standard of living.

Crop production is the third or fourth most important source of income for most households – following livestock tending (and sales), wage employment and beer brewing. Remittances from family members working outside the village also tend to be a more important source of income than crop production, at least for poor households. During drought periods, transfers from the government have provided more real income for many households than has crop production.

While out-of-village wage employment is very important, in-village, off-farm income opportunities are poorly developed. Traditional marketing systems, which provide many employment opportunities in West Africa, are virtually non-existent in Botswana. The reasons for this are not clear. One major factor may be that there has been no 'engine of growth' in Botswana to foster its development in the form of crops produced for disposal. Participation in the Southern African Customs Union has also restricted opportunities for the development of rural small-scale industries. It should be emphasized that current management systems represent reasonable responses to the realities of the natural environments and socioeconomic circumstances.

Local trade historically has been dominated by informal exchanges between households. Informal trades are made for labour, traction, veld products and household items on both a cash and barter basis. Interhousehold exchanges significantly increase the access of poor farming families to required production resources.

Complementing household exchanges, there is in Botswana a network of privately owned general traders, which extends to essentially all villages. The general traders distribute imported agricultural commodities and other household goods to rural areas. Most of the items sold are either processed or manufactured and are of (nearly) standardized quality. Prices are subject to a price control system based on wholesale prices and transport costs. Seed for sorghum and other crops is available through the Botswana Agricultural Marketing Board (BAMB) depots located in the larger villages. BAMB also announces guaranteed producer prices before the beginning of each season. In addition, the availability and low prices of imported, milled food grains clearly affect farmers' incentives to invest in sorghum production.

Farming families in Botswana have systematically tried to minimize the linkage between food production and food security. While the main objective in growing sorghum is to meet household food grain requirements (Baker, 1987), no households actually rely on food self-sufficiency. Instead, most food is purchased using cash from a combination of several sources. If the rains are good and a sufficient quantity of sorghum is produced, it is a welcome event. The money, which would have been spent on food, becomes available for other items. The food security strategy of farmers implies that capital or labour investments in sorghum production usually are not a top household priority. This attitude poses a major challenge to technology development researchers.

National resource endowments can also affect farmers, as recent experience has demonstrated. Government revenues from cattle and diamond exports have been used to subsidize farming families – substantially mitigating the most serious effects of droughts of the 1980s and 1990s.

Strategies for sustainable rural development

The foregoing discussion has indicated that many factors, both biophysical and socioeconomic, influence rural livelihoods. However, it is also obvious that in the drought prone areas water becomes one of the most critical determinants of the level of that livelihood. Given that the potential for irrigation is rather limited in much of the region, it is obviously important to have a good understanding of the rainfall patterns in designing and implementing strategies for improving the level and reliability of rural livelihoods (Nicholson, 1986). These strategies can be considered in terms of the design/development and dissemination/implementation of relevant improved technologies and policy/support systems.

Since the sustainability of Botswana's farming systems has become questionable because of the influence of outside factors, the responsibility for improving agricultural productivity must fall mainly on agents other than farmers. However, the probability of success is likely to be much higher if two simple lessons learned from the past are taken into consideration in designing potential improvements. These are to consider both rainfall variability and traditional farmers' wisdom in such technologies.

Consideration of rainfall variability

The agricultural technology developmental process must start from the premise that water availability is a variable rather than a constant. In designing relevant technologies/strategies to deal with this, agricultural researchers should collaborate with climatologists to obtain a good understanding of the level, intensity and patterns of rainfall of the area(s) where they are working. As well as knowing the probability of specific amounts of rainfall by, for example, 10-day intervals throughout the year, to know whether, and for how long, rainfall is likely to exceed potential evapotranspiration is also valuable. Based on our experiences in northern Nigeria and Botswana we would argue that such knowledge would result in very different approaches to developing relevant improved technologies.

For example in West Africa, as one moves north through the savannah into the Sahelian region, annual rainfall becomes lower while the rainy season becomes shorter and more variable in terms of the dates of onset and finish of the rains. However, even quite far north there is a period of two or more months when rainfall exceeds potential evapotranspiration with a reasonable degree of certainty. This period provides a window of opportunity for crop researchers to develop technologies (e.g., short cycle crops) that can respond to the shorter season.

In the semiarid areas of Botswana, however, the rainfall pattern is very different with no period when rainfall exceeds potential evapotranspiration with any degree of regularity. Such an environment would seem more suitable for livestock, but as in West Africa, because of their relatively poor economic situation, many rural households depend on growing crops as their major source of income. Here, the challenges for crop researchers in developing relevant improved technologies become even greater. In such a situation, it is critically important that the major crops have a great deal of *plasticity* — that is, be able to sit and survive during drought periods and be ready to respond rapidly to soil moisture when it becomes available. Therefore crops with such characteristics may be rather long season in duration (e.g., the traditional sorghum variety Segaolane in Botswana). There is, however, still some role for short cycle crops (i.e., the shorter the better), particularly if planting rains come very late in the season, and for *minor* crops (e.g., cowpeas) which can take advantage of periods when the rains are favourable.

Another rainfall variable that should be considered is the distribution and intensity of rainfall, particularly at the beginning, but also, to some extent, during the growing season. This information would be useful in determining the value as developing technologies that support water harvesting (e.g., planting in dips, tied ridging, etc.). In Botswana, for example, the window in terms of days for planting on optimal soil moisture was increased through double ploughing, with the first ploughing being done on days when there was insufficient soil moisture for planting (Heinrich *et al.*, 1990). The rationale was that improving the permeability of the soil through the first plowing allowed any rainfall to infiltrate the soil and to be available for planting when the next rainfall fell. However, intensity of rain during the growing season also needs to be known too much at once could have a negative impact on the water harvesting techniques by causing water logging and, in some circumstances, erosion.

Consideration of farmers' wisdom

Although the strategy just described — designing technologies to respond to the variability in rainfall — is a useful starting point, it is still likely to be imperfect, particularly when the rainfall pattern is difficult to predict with any degree of certainty. Nevertheless, earlier farmers have managed to survive and, to some extent, flourish in uncertain climatic environments.

Given the fact that farmers are natural experimenters and that tradition often is based on what has been found to work best in the past and also that they have an intimate knowledge of their *total* environment, researchers should harness their expertise in developing relevant technologies that have a strong scientific base. Fortunately, the need for this has been recognized increasingly in recent years with the popularization of farmer participatory methodologies and the farming systems approach to research and to extension. This approach has now been institutionalized in Botswana. Developing relevant improved technologies constitutes only one part of the puzzle of improving rural livelihoods. The policy/institutional support system is also important.

Establishing a relevant policy/institutional support system

The following components need to be considered in developing and implementing relevant policy/support systems.

Ensure congruence between technology and the policy/support system. This relationship is obvious and requires little further discussion. However, perhaps one area that warrants specific attention is the need for governments to be more interventionist in supporting the development of the private sector to shoulder some of the responsibility for services previously provided by the public sector. However, development of private sector interest is unlikely unless governments can ensure that the physical infrastructure (e.g., roads) and information dissemination system (e.g., prices) are well developed, and that a facilitative regulatory and institutional environment is in operation.

Accept drought as normal part of cycle and act accordingly. Botswana is an example of a country where the potential for drought occurring was explicitly taken into account when drawing up the 199197 Development Plan. In cases similar to Botswana, explicit recognition of the endemic nature of drought provides the potential for constructively using drought periods to implement strategies that will improve future agricultural productivity while at the same time ensuring sustainability.

The interdependencies between drought relief, production, and sustainability become even more critically important the longer the drought periods are. In southern Africa, for example, one researcher (Tyson, 1979) identified a rainfall cycle consisting of 10 years above average rainfall and 10 years below average rainfall. Farming households that operate very close to the survival level in normal years becoming increasingly vulnerable the longer the drought periods persist. To eliminate the devastating impact of drought periods on the livelihoods of rural families, three initiatives are required:

1) Some form of rainfall monitoring system should be put in place. Short field visits might supplement the system to provide an early warning of impending problems in terms of production shortfalls on the part of the farmers. One source of information, which may be useful, is the Famine Early Warning System bulletin, which is financed by USAID. We are not sure, however, whether such information can be desegregated sufficiently for the purposes we are advocating.

2) A timely response must be made to potential problems that have been identified. Although farmers have over time developed strategies to cope with

such negative situations, their ability to respond is limited, particularly if the drought period is prolonged. The usual pattern that develops once opportunities for current work are exhausted is to sell consumption assets, followed by productive assets, and then to migrate as a last resort (Frankenberger and Hutchinson, 1991). Once productive assets such as animal traction and seed stocks start being sold the road back to full production is much more difficult, often requiring considerable external assistance. What is required is to provide help to such families before sales of productive assets start, thus paving the way for rapid recovery after the drought period.

3) Innovative ways must be found to give help that can be put to constructive use. The immediate issue is, of course, to ensure that the problems of hunger resulting from drought are addressed by distributing food to the needy. However, whenever possible, it is desirable for such food to be distributed through food (or money) for work programs which ensure strategies are put in place that improve potential productivity and sustainability.

4) Botswana currently provides a good example of the way in which such a program can be implemented. The Botswana government now, according to their Development Plan (Sigwele and Norman, 1993), maintains some flexibility when certain development projects, particularly those designed to improve infrastructure, conservation or sustainability, are implemented. Implementation of these is quickly accelerated during drought periods, thereby providing work for people. This Labour Based Drought Relief (LBDR) Programme pays about 50% of the minimum daily wage thereby discouraging all but the most needy from participating in it. Village Development Committees (VDCs) have primary responsibility for local implementation of the LBDR initiative. They select the workers and the projects from a portfolio of development projects drawn up earlier.

Ensure complementarity between productivity and ecological sustainability

In the drought prone areas of Africa, strategies designed to enhance ecological sustainability must in essence be 'piggybacked' on those designed to encourage improvement in short-run productivity. Thus a convergence is required between the short-run interest of rural households in producing enough food to last until the next harvest season and the long-run interest of society of ensuring the sustainability of the land so that it can be used by future generations. Strategies that can help to ensure that such synergism occurs include the following (Norman and Douglas, 1994):

1) *Technological thrust* Researchers need to place greater emphasis on *ex ante* screening of all potential technologies, to ensure that their adoption will have no negative environmental impact, and preferably go one step further to ensure that they will not only have a positive production impact but also contribute positively to soil productivity.

2) *Policy thrust* Very often policies designed to improve productivity are in conflict with those designed to encourage sustainability. For example, a subsidy is given for destumping fields to facilitate ploughing but no compensation is offered to encourage living hedges to prevent erosion. Incentives may encourage food self-sufficiency in the major food crop while discouraging diversification so important in facilitating sustainability. Two obvious components of the policy thrust are to eliminate possible conflicts between policies designed to promote short-run production and those designed to encourage long-run sustainability, and to positively encourage, via incentives and local responsibility, adoption of strategies that conserve the environment for use by future generations.

3) *Corrective measures* The physical structure that refer to the protection of the soil.

The above two thrusts in essence constitute preventative strategies; whereas, corrective measures conform to the more traditional approach to soil conservation. Here, emphasis is placed on physical structures to prevent further erosion once it has developed or to prevent new erosion on land just being opened up to cultivation. Appropriate solutions need to be based on an understanding of the causes of soil erosion, not just the symptoms.

Obviously the breakdown of strategies is somewhat simplistic since the three thrusts are not necessarily mutually exclusive. For example, policies can help or hinder the use of structures. Consequently, a combination of strategies is likely to be needed to maximize the effectiveness of soil conservation programs.

Return of some responsibility to the community level

There is, in many countries, a trend towards handing back some control and responsibility to local communities, particularly with respect to natural resources such as communal grazing land and wildlife development strategies. These policies have merit because local communities have a much better idea of what will or will not work in the local setting and will reap the benefits of strategies implemented to ensure continued productivity of communal resources.

Initiate population control and development of the non-agricultural sector

It would be naive to believe that improving rural livelihood through simply targeting initiatives directly to them will be sufficient. The high population growth rate coupled with the fact that agriculture is the "residual employer of people" means that strategies that go way beyond the focus of this paper need to be addressed. Examples of these are to introduce strategies designed to reduce population growth rates, to develop employment opportunities in the non-agricultural sector, and to promote part-time off-farm employment in rural areas. Unfortunately these strategies are not particularly easy to implement especially when the majority of the population is still engaged in agriculture. Also, unfortunately, supplementary employment in the rural areas is often associated with, or a spin-off from, the commercialization of agriculture. However, success in developing this type of employment opportunity can help in relieving the ever-increasing pressures on the natural resource base. Nevertheless, unless success is achieved in implementing these strategies, it is unrealistic to expect improvement of rural livelihoods in a sustainable manner.

Conclusion

In the drought prone areas of Africa, to recognize that agriculture holds the key to alleviating poverty in the foreseeable future is important. Also, given the close link between poverty and environmental degradation (Mellor, 1988; Pinstrup-Anderson and Pandy-a-Lorch, 1994; 1995), the latter will not be solved without first addressing the former – a formidable task, especially given the apparent decrease in donor and national support for agricultural research and many developmental initiatives (Oram, 1995; Pinstrup-Anderson and Pandy-a-Lorch, 1994). The ideal strategy for rural development in marginal drought prone areas is one that not only is compatible with existing systems of production but that also builds upon the strengths and potentialities of the existing systems.

References

Agence France Presse (1993), Soil loss could make vast lands infertile, provoke famine: report, *Agence France Presse.*

CGIAR (1992), *Review of CGIAR priorities and strategies Part*, CGIAR, Washington.

CGIAR (1994), *Research for a food secure world: an overview*, CGIAR, Washington.

FAO (1987), *Improving productivity of dryland areas*, Committee on Agriculture Report 87/7, FAO, Rome.

Frankenberger, T.R., and C.F. Hutchinson (1991), *Sustainable resource management based on a decentralized food security monitoring system*, Discussion Paper, Office of Arid Lands Studies, University of Arizona: Tucson, USA.

Heinrich, G., J. Siebert, E. Modiakgotla, D. Norman and J. Ware-Snyder (eds) (1990), *Technical summary of ATIP's activities, 1982-90: research results*, ATIP-1991-RP-5, Department of Agricultural Research, Ministry of Agriculture, Gaborone, Botswana.

IITA (1994), *Medium term plan, 1994-1998*, IITA, Ibadan.

Mellow, J.W. (1988), The intertwining of environmental problems and poverty, *Environment* 30(9):830.

Nicholson, S.E. (1986), Climate, drought, and famine in Africa, in Hansen, A. and D.E. McMillan (eds.), *Food in Sub-Saharan Africa*, Lynne Rienner, Boulder, CO, pp. 107-128.

Norman, D.W., and M. Douglas (1994), *Farming systems development and soil conservation*, FAO Farm Systems Management Series Number 7, AGSP, Food and Agricultural Organisation of the United Nations, Rome.

Oram, P. (1995), *The potential of technology to meet world food needs in 2020*, 2020 Brief No. 13, International Food Policy Research Institute, Washington.

Pinstrup-Anderson, P. and R. Pandy-a-Lorch (1994), *Alleviating poverty, intensifying agriculture, and effectively managing natural resources*, Food, Agriculture, and the Environment Discussion Paper No. 1, International Food Policy Research Institute, Washington.

Rwegayura, A. (1992), Environment: Africa urges new partnership for development, *Inter-Press.*

Sigwele, H. and D. Norman (1993), Rural development in Botswana: a case study. Invited for *Agriculture: Food, Environment, and Rural Development Conference, Salzburg Seminar*, sponsored by the Kellogg Foundation, Schloss Leopoldskron, Salzburg, Austria, October 30th - November 6th.

Tyson, P.D. (1979), Southern African rainfall: past, present and future, In Botswana Society (ed.), *Symposium on Drought in Botswana*. Clark University Press, Worcester, USA.

World Bank (1995), *The many faces of poverty in Africa*, Findings Number 35, Africa Region, World Bank, Washington.

12 Global change and community self-reliance strategies in Southern Africa

ETIENNE NEL

Introduction

The 1980s and 1990s have witnessed tremendous changes in the global economy and the position of countries relative to each-other in terms of their ability to compete effectively in an increasingly globalized economy. Whilst some areas have benefited, the picture is not a uniform one. Debt, structural adjustment programmes and the reduced role of the state in development intervention have had serious ramifications for the lives of ordinary citizens in Africa, the world's poorest continent. The lowering of global trade barriers has witnessed the disappearance of inefficient producers and led to job losses in many areas. Amongst the hardest hit areas in Africa have been those which are economically and environmentally marginal. Despite this bleak picture, as is not uncommon in situations of extreme adversity, human ingenuity and resourcefulness often triumph as a result of people's desperate attempts to survive. A net result has been the search for alternate forms of income and livelihood in such rural areas, which rely on either traditional skills or adapted, appropriate technology and techniques. The pursuance of new forms of employment, a frequent return to traditional activities and attempts to ensure food security has been variously labelled as, 'development from below' and 'community-based, self-reliance' (Gooneratne and Mbilinyi, 1992; Taylor and Mackenzie, 1992; Burkey, 1993). This paper explores these issues, with specific reference to marginal rural areas in the Southern African countries of South Africa and Zimbabwe, with a view to identifying the type of community self-reliance responses adopted and their sustainability. Case studies drawn on are used as cameos to illustrate the varying responses adopted in impoverished rural areas to the development challenges faced, the relative degrees of success attained and the role played by external agencies. It should be noted from the outset that the two cases investigated, are examples of relatively successful endeavours and as such reflect the potential rather then the actuality of such endeavours in the broader region.

Africa in Crisis?

Western notions of the 'global village' ignore the inability of the majority of the world's population to adequately participate in such post-modern concepts, for as Esteva and Prakash assert, '...far from being "globalized", the real lives of most people on Earth are clearly *marginalized* from any "global" way of life. The social majorities of the world will never, now or in the future, have access to these so-called global phenomena' (Esteva and Prakash, 1997, p.285). Desperate economic realities in Africa compel many communities to seek their own solutions to the circumstances in which they find themselves. According to the United Nations, '... (for) approximately one-sixth of mankind, the march of human progress has become a retreat. In many nations, development is being thrown into reverse' (Gooneratne and Mbilinyi, 1992, p.22). The precarious economic situation in post-independence Africa was dealt a cataclysmic blow through a combinations of the 1970s economic and oil crises and the 1980s debt crisis. Africa's failure to meet the demands of western countries and bankers led to the imposition of structural adjustment packages by the Bretton Woods institutions in the 1980s. Standard, imposed measures such as currency devaluation, withdrawal of the state from many facets of the economy, rationalization of the civil service, the curtailment of food and other subsidies have frequently had far-reaching results, helping to contribute to crises ranging from bread riots in Zambia, to inflation rates exceeding 1000% in the Congo and the retrenchment of nearly half of the civil service in countries such as Ghana and Uganda (Barratt Brown, 1995). According to Taylor and Mackenzie (1992), in countries subjected to harsh structural adjustment packages, their economies contracted by 0.5% p.a. in the 1980s. Although measure of success were attained in some countries such as Uganda, the overall picture is extremely negative. By 1990, Africa's exports of money to meet debt repayment requirements equalled its earnings from commodity exports.

The 1980s was a bleak decade for Africa, average income declined by 2.6% p.a., wage employment fell by 16% and the GDP p.c. fell from $ 854 in 1978 to $ 565 in 1988 (Taylor and Mackenzie, 1992, p. 215). Although there has been a partially turn around in fortunes in the 1990s with some countries returning to positive growth economies, there must be seen as against the reality of a very low economic base. In countries such as Zambia the currency is now valued at less than a thousandth of what its international value had been in the 1960s, in Zimbabwe the current inflation rate has exceeded 50% whilst in South Africa the unemployment rate is of a similar magnitude.

These issues are felt particularly acutely in rural areas which have suffered the combined effects of static or falling commodity prices, contracting markets as a result of foreign competition in the global economy, physical isolation, government's which are unwilling or unable to offer any meaningful support and natural disasters, such as the 1970's Sahel and the 1990s southern African droughts. Making ends meet in such areas has become a constant battle and what Taylor and Mackenzie (1992, p.1) refer to as 'the struggle to survive'.

New hope for Africa?: The move to self-reliance

Despite this extremely negative picture it would be wrong to assume that depravation and marginalization are the order of the day. There are signs, at the community level, that there people and communities with drive who are prepared to tackle their unique problems, with or without external help. These stirrings of self-reliance have not gone unnoticed in the media. A prime example is Time magazine which in a special feature on 30, March 1998 (pp. 39-40) noted:

> ....'out of sight of our narrow focus on disaster, another Africa is rising, an Africa that works...What's new is how some nations are figuring out ways to harness their natural and human resources into working models of development, even while others cannot. What is new is the astonishing extent to which ordinary Africans are searching out their own paths to progress. What's new is how much of the still limited prosperity and security they have managed to acquire is homegrown - political and economic advances rooted in the soil of local culture.'

Though limited in scale, these are encouraging signs which offer the chance of socioeconomic improvement to members of the world's poorest societies and which find accord with recent writings on self-reliance and 'bottom up' development.

Development in Africa is increasingly recognizing and encouraging local initiative and the positive role which aid agencies and Non-Governmental Organizations (NGOs) can play. This is partially as a result of the perceived 'impasse in (conventional) development theory' (Schuurman, 1993). Observed changes in community-based development over the last twenty years are belatedly being acknowledged and incorporated into theoretical constructs (Pieterse, 1998). The early, and perhaps somewhat limited, principles of 'development from below' and 'development from within' are being drawn together with empirical evidence into a more sustained framework by the recent emergence of the 'alternate development' (Hettne, 1995) and the 'anti-development' paradigms (Stock, 1995). It seems that these discourses originate in the current crisis of development and from the achievements of grassroots groups. Escobar argues the potential of '... the search for alternate ways of organizing societies and economies, of satisfying needs, of healing and living' (Escobar, 1995, p.226). Even though such ideas have come in for legitimate criticism recently (Pieterse, 1998), as Pieterse (1998) convincingly argues, through a dialectical process, mainstream development has gradual become more reflexive, selectively adopting new ideas and modifying its focus to acknowledge and accept a more participatory element.

The switch in applied development theory, policy and practice to a allow for a more 'bottom-up' approach, accords with Stöhr's (1981) seminal argument that the failure of modernization approaches to tackle poverty requires new forms of intervention in which control is vested in the host community. Communities need to draw on local skills, resources and indigenous knowledge. Research undertaken in various parts of Africa reveals that economic marginalization is encouraging a new or enhanced focus by communities and aid agencies on the development of local potential through indigenous strategies (Taylor and Mackenzie, 1992). The principle for 'self-reliance' is inherent within this switch in focus, a concept which finds accord in the writings of the South Commission which has come out in support of a 'revival of economic growth which is inward-orientated and self-sustaining, providing full-employment and the development of both agriculture and industry' (Gooneratne and Mbilinyi,

1992, p. 3). It is however important to note that in order for independent initiatives to succeed, human and social capital must be of a high level. In situations where both are poorly developed, development prospects will be commensurately restricted.

Even though the principle of community-based self-reliance is based on indigenous control and initiative, the role of external 'change agents' should not be ignored. According to Burkey (1993, p. 73), 'self reliant participatory development processes normally requires an external catalyst to facilitate the start of the process and to support the growth of the process in its early phases'. Within this context, it is argued that even if governments lack the resources to assist marginalized communities directly, creating the supporting economic environment can be critical. In addition, there is increasing recognition of the potential and actual role of non-governmental organizations (NGOs) and external donor agencies in helping to initiate and support self-reliance (Burkey, 1993). 'While external inputs may be considered contradictory to self-reliance, carefully chosen external support may be useful to overcome severe resource constraints, and provide a protective political umbrella' (Gooneratne and Mbilnyi, 1992, p. 265). Naturally, support given should be of a short-term nature and it should not suppress local initiative, skill or control.

Although Stöhr's (1981) concept of 'bottom-up' development has much to commend it from idealistic and practical perspectives, as Stock (1995) points out, such action is unlikely to be successful beyond a limited handful of cases. In fact, according to both Stöhr (1981) and Simon (1990), 'bottom-up' initiatives need to complement and harmonize with 'top-down' support and facilitation. Whilst such an argument is commendable, the practical reality is that government in he South is frequently impoverished and incapable of rendering realistic support. External support therefore often tends to be dominated by the NGO sector. In addition, even in remote rural villages, the 'global village' has extended its tentacles and the success or failure of community self-reliance might well be determined by external, market forces. Such forces can either facilitate or hinder community endeavours, dependent on the initiative's comparative advantages and whether or not it poses a competitive threat to other market role-players. In this paper the important role which external 'change' agents and outside agencies can play will be highlighted in the cases examined.

The case studies

In the following sections, examples of community-base self-reliance in Zimbabwe and South Africa are examined to illustrate the vibrant nature of local initiative and resourcefulness which exists, but similarly they also show the practical and market related constraints which frequently influence them. The potentially important role of external aid agencies is an important additional theme in the basic argument.

A case study of self reliance through beekeeping in Bondolfi, Zimbabwe

Economic marginalization, structural adjustment and the devastating drought which affected much of southern Africa in the early 1990s have had a severe impact on the economy and society of Zimbabwe. In that country's eastern and southern regions, the semi-arid nature of the environment has further exacerbated the situation through restricting opportunities for agricultural diversification, limiting the region's economic and social development (Matizha, pers. com., 1998; Sister Chiendsa, pers. com., 1998). In this part of Zimbabwe, poor soils and drought frequently make the securance of external sources of income, besides conventional agriculture, essential. The situation is aggravated by high population densities, which places further pressure on the already limited natural resources. Households in this part of the country are therefore subject to '...widespread food insecurity resulting in deleterious impacts on health' (Scoones *et al.*, 1996 p. 11). It is as a result of these social and economic changes that innovative responses are called for on the part of rural households in order to make ends meet financially. In this case-study of the Bondolfi area in Zimbabwe, the contribution which beekeeping is making to help address the situation which prevails is examined. This study also indicates the important and positive support which well intentioned NGOs can render to impoverished communities in times of crises.

The Bondolfi area consists of a series of villages scattered across an area of bush, low hills and agricultural fields, some 40 km. south of the district centre of Masvingo. The most prominent nodes in the area are a few stores, a Roman Catholic Teacher's College, Church and School and a Mission. The area is currently occupied by some 10,000 people, almost all of whom depend on small-scale farming for their food supply and income (Matizha, pers. com., 1998). The main crops grown are cotton, maize, sorghum and groundnuts (Zvovashe, pers. com., 1998). In Bondolfi, beekeeping has always been pursued as a traditional activity, but has seldom been a primary source of income (Mutude, pers. com., 1997), being used instead as a secondary source of food, for barter and as a traditional medicine. Medicinal uses of honey in the community include the treatment of wounds, its use by diabetics, for feeding infants and to treat malnutrition, anaemia and scurvy (Matizha, pers. com., 1998).

In recent years members of the community, led by local leaders with a vision for change, in collaboration with the church, realized that enhanced and improved beekeeping and honey production could play a significantly more prominent role in the area. To their credit the community realized that a measure of outside support was needed to maximize local potential and earnings from beekeeping. Fortuitously, this coincided with the interest shown by an external aid agency, the United Nations Volunteer program in the area, the positive support which they were able to provide and material assistance received from the European Union (Chale, pers. com., 1997). Based on the former's advice, improved beekeeping skills and marketing strategies were acquired (Mutude pers. com., 1997). One of the first actions of the community members was to establish the Bondolfi Beekeeper's Association in 1995 to co-ordinate activities and to undertake the bulk-collection and processing of honey. The Association now (1999) has 70 active members, 40 of whom are women. The endeavour is assisting women, who are traditionally disempowered in Zimbabwean society, to acquire an independent livelihood. On average each member manages two hives, although it is the intention of the United Nations volunteer to increase the number of hives, through time, to at least ten per member (Chale, pers. com., 1997). The production of new hives has, in itself, generated local employment opportunities for carpenters who manufacture hives at the association's small shop in the area. Honey sales have increased several-fold and the income of participating families by a factor of nearly four since the project was initiated. Other direct spin-offs include the establishment a bee-hive smoker enterprise in Bondolfi, and with the help of the church, a leather processing industry which manufactures leather gloves for beekeepers and a sewing group which produces veils for the collectors has been established (Chale, pers. com., 1997).

The association also co-ordinates the collection, processing, marketing and sale of honey and wax, thus overcoming the extreme difficulty which individual producers previously experienced in terms of trying to sell their produce. The association has overcome the constraints which individuals had encountered in trying to get their goods to Masvingo, the closest major urban centre. Honey is now collected and bottled centrally and then sold in bulk, in jars carrying the association's logo. In addition to obvious savings in time and effort, the sale of a product which has become recognized for its quality is able to command higher prices and turnover than individual producers could have hoped to achieve (Mutude pers. com., 1997). Increased sales have resulted, which in turn has had a cumulative effect in Bondolfi where income and employment has risen. Although it is still too early to see noticeable material improvements in Bondolfi, the attainment of a more reliable form of income holds considerable long-term benefits for the community. The role of the church and the UN volunteer in helping to attain the vision and to provide skills training, transport, advice and technology should not be under estimated. Without such input, improved production techniques would not have been realized and the urban market would not have been penetrated.

The Mansomani initiative

The second initiative examined is the agricultural co-operative in the Mansomani area in South Africa. In this case, community initiative and vision, similar to that displayed in the last case is also evident. The scheme however differs in that this initiative secured a market ally in the form of an establish agri-business which was prepared to invest in the area and combine its own interests with those of the community for the mutual benefit of both. In a very real sense this project would appear to be succeeding because the external agency has helped to address the local capacity and technology shortfalls and, equally importantly, has guaranteed a market share for the produce of the community initiative. At another level, instead of falling victim to the global system, Mansomani has succeeded in participating within it. As South Africa is one of the world's largest exporters of sugar, agri-business is always looking for new production fields, and in the case of Mansomani a business relationship which benefits both partners was negotiated.

The Context   In the deep rural areas where there are almost no formal employment opportunities, local communities are forced to identify and implement appropriate self-reliance strategies in order to survive and earn some form of an income. Interestingly enough, unlike in many other parts of South Africa, a system of freehold tenure does not prevail here. Instead, land is held under tribal / communal tenure under the headmanship of a local chief in this area (Botha, pers. com., 1997). In the area's favour is the fact that it is well-watered and it also has fertile valley soils. A high percentage of the country's sugar-cane crop is sourced from the coastal and adjacent plains of the KwaZulu-Natal province in which Mansomani lies. The use of almost all available commercial land means that agri-business is anxious to gain inroads into suitable lands held under tribal tenure. However, because of the inflexibility of the tribal tenure system and the distrust of chiefs, there are however only limited cases where this has actually happened. The case of Mansomani is one in which a community, which possesses access to their land under communal tenure, has entered into a business contract to sell sugar-cane to a nearby sugar mill in return for advice, extension support and a guaranteed market. Mansomani therefore differs considerably from the other initiative outlined in this paper, in that whilst Bondolfi has entered the commercial market, it has been on an independent and low-key basis. In Mansomani the community have participated directly with large commercial enterprise and operate through a direct partnership with a key agro-industry (Khuzwayo, pers. com., 1997).

Despite the patent advantages which could exist in ecologically suitable areas for communities through such arrangements, only a handful have taken advantage of such relationships. As is not uncommon in such endeavours, the case of Mansomani differs owing to the presence of a gifted community leader, Victoria Khuzwayo, who was successfully able to mobilize community support for her development proposals. In addition, this local champion won the support of the area's chief and hence gained access to land. She also helped to forge the business link with the local sugar-mill (Botha, pers. com., 1997). This all took place despite generations of discrimination against Blacks and Black women in particular, who were denied opportunities in a society which has traditionally been dominated by racial and gender prejudice.

The Development Initiative In 1982 the chief granted 65, mainly women farmers, the right to clear some 200 ha. of bush. Following an agreement with the local sugar mill, with whose help and earth-moving equipment, the area was converted to irrigated sugar-cane fields, the small farmers were able to commence farming with the knowledge that there was a secure and reliable market for their output. In addition, the mill installed two pumping stations and employs a permanent staff to maintain the irrigation system and advise the farmers. Farming activities are undertaken on a communal basis with the mill supplying extension services and cane cuttings for new plants (Mxeke, pers. com., 1997). The community now produces over 18, 000 tons of cane a year. The clear financial and other gains from its sale by the community are obvious, albeit that they have no choice over what crop they grow. In addition to what, for many families, have been the first significant incomes which they have earned, the community has been able to build a community hall, a school, a crèche for pre-school children and a brick-making plant with their profits. There have been improvements in the overall quality of life in the district and the success of the project has given people a significant sense of self-achievement (Khuzwayo, pers. com., 1997). Despite the occasional vagaries of drought and price fluctuations, the Mansomani initiative appears to have gone from strength to strength over the last 17 years. The leadership structure is intact, tangible gains have been made in terms of income secured and community facilities, loans have been repaid and the guaranteed market access have ensured the project's on-going success (Khuzwayo, pers. com., 1997).

Assessment and conclusion

Although there are clear similarities between the various initiatives, in terms of their experience of disempowerment, and also their attempts to strive for community-based self-reliance, clear differences also exist. Key points of similarity include the fact that, in both cases, community-based strategies, motivated by prevailing economic crises and the presence of talented leaders helped to catalyze the projects. In Bondolfi the combination of local skills and resources and external advice and support has led to a successful, empowering initiative, which has improved the economic conditions of participants considerably. The scheme has not however transformed the community in the same sense that the Mansomani initiative has, Bondolfi has only secured a very limited stake in the free market and has ensured survival not prosperity. In the case of Mansomani, the community could offer a resource which was in high demand to local capitalist enterprise, namely access to land. The combination of local leadership and talent and external advice and support has helped to secure guaranteed market access which ensures that both partners benefit. Unlike the other initiative Mansomani has moved beyond mere survival levels, and in a small way, has been effectively integrated into the dynamics of the global market and has not fallen out of the market system as is so common with other community initiatives in Africa (Gooneratne and Mbilinyi, 1992). On a positive note, It is significant to note that the strategies are also addressing the gender issue, by empowering women who have been traditionally marginalized.

Even though these cameos cannot provide insight into the totality of rural development in southern Africa, they do serve to illustrate salient points which are helping or hindering rural development initiatives in marginal areas. Self-reliance / 'bottom-up' development in response to very real environmental and economic challenges in marginal areas are occurring. The case studies indicate that local vision, leadership and drive does exist in rural communities to improve conditions. However, in a global order increasingly dominated by free-market logic and free trade, there is an ever diminishing space for those lacking in skills and resources to compete effectively. Whilst not seeking to moralize about this situation, given the near collapse of many governments in Africa in the face of internal difficulties and changes imposed by the global economy and institutions, the picture for disempowered communities who either do not have a key asset to market or unique local skills or the support of a sympathetic external agency or business partner does seem bleak indeed. If a community does not have access to a defined resource such as land, or does not produce a particular crop for which a niche market exists, there is little room for what, in relative terms, would otherwise be marginal level participation in the commercial market. If the commercial market is difficult to penetrate for community-based agriculture, such initiatives that do exist have little scope to go beyond meeting the subsistence needs of the areas in which they operate. Unless market shares can be assured, or business / government adopts a defined policy of buying from community farmers, emerging Black commercial farmers are probably going to remain in a marginalized position.

Local and international donor agencies do exist, which are prepared to support community initiatives such as that in Bondolfi. However, few of these rise much beyond survival levels. What Mansomani illustrates is the rare case of a local community which has succeeded in entering the free market, based on its ability to provide a resource desired but legally unobtainable to the free market. Indications are that in southern Africa the prospects for similar initiatives are limited and that the agricultural sector will continue to be dominated by the large, commercial producers and agri-business. Another key theme which the studies reveal is the valuable role which external agencies can and are playing, be they the NGOs or the church in Bondolfi or private business in Mansomani. Such agencies have the ability to train communities to better manage their resources, the provide market access and technical support. What these studies reveal is that as Stöhr (1981) and Simon (1990), suggest 'bottom-up' initiatives need to complement and harmonize with 'top-down' support and facilitation in order to ensure success. In principle, government must also play a key role play, particularly in terms of extension support and the facilitation of initiatives, however their own capacity and funding constraints are a stumbling block in this regard.

Whilst it is recognized that the experiences detailed do contain certain unique elements, they do reinforce the argument, namely that there is scope for a new paradigm of development which draws upon local experience and does not rely on imposed goals and practices. In order to succeed, such approaches need to be rooted in the culture, ideals and mind-sets of those groups engaged in the development process. Ideally, development should foster self-reliance and community-based initiative, and as Escobar suggests, '...this can best be achieved by building upon the practices of the social movements ...(which) are essential to the creation of alternate visions of democracy, economy and society' (Escobar, 1995, p.212). However, as this article shows, situationally relevant development also requires appropriate external facilitation and support, access to market and exploitable local resources if it is to succeed. The experience derived from case-studies such as these suggests that researchers should not reject the concept of development as some writers would encourage, but rather we should strive, as Simon (1997) suggests, to formulate 'different paths' to achieve the same goal. This concurs with calls made by writers such as Edwards (1993), that development research should be 'relevant' and contribute meaningfully to new conceptualizations of development.

References

Barratt Brown, M. (1995), *Africa's Choices: After Thirty Years of the World Bank*, Penguin, London.

Botha, T. (1997), Personal communication, South African Sugar Association, Durban.

Burkey, S. (1993), *People First*, Zed Books, London.

Chale, T. (1997), *Summary of the Bondolfi Beekeepers Association*, Faxed note, Bondolfi.

Chiendza, Sister (1998), Personal communication, Swiss Mission, Bondolfi.

Edwards, M. (1993), 'How relevant is development studies?', in, F.J. Schuürman (ed.), 1993, *Beyond the Impasse: New Directions in Development Theory*, Zed Books, London, pp. 77-92.

Escobar, A. (1995), 'Imaging a Post-Development Era', in J. Crush (ed.), *Power of Development*, Routledge, London, pp. 211-227.

Esteva, G, and Prakash, M.S. (1997), 'From global thinking to local thinking', in M Rahnema and V Bawtree (eds), *The Post-Development Reader*, London, Zed Books, pp. 277-289.

Gooneratne, W. and Mbilinyi, M. (eds), 1992, *Reviving Local Self -Reliance*, United Nations Centre for Regional Development, Nagoya.

Hettne, B. (1995), *Development Theory and the Three Worlds*, Longmans, Harlow.

Khuzwayo, V. (1997), Personal communication, community leader, Mansomani.

Matizha, H. (1998), Personal communication, member of the Bondolfi Beekeepers Association and a shopkeeper.

Mutude, J.M. (1997), Personal communication, ex-Secretary Bondolfi Beekeepers Association.

Mxeke, W. (1997), *Personal communication*, Glendale Mill, Mansomani.

Pieterse, J.N. (1998), 'My paradigm or yours? Alternate development, post-development, reflexive development', *Development and Change* 29, pp. 343-373.

Schuurman, F.J. (ed.) (1993), *Beyond the Impasse: New Directions in Development Theory*, Zed Books, London.

Scoones, I. (1996), *Hazards and Opportunities: Farming Livelihoods in Dryland Africa: lessons from Zimbabwe*, Zed Books, London.

Simon, D. (ed.) (1990), *Third World Regional Development: A Reappraisal*, Paul Chapman, London.

Simon, D. (1997), *Development Reconsidered: New Developments in Development Thinking*, paper presented at the Society of South African Geographers' Conference, Midrand.

Stock, R. (1995), *Africa South of the Sahara: A Geographical Interpretation*, Guilford Press, New York.

Stöhr, W.B. (1981), 'Development from below: the bottom-up and periphery-inward development paradigm', in W.B. Stöhr, and D.R. Fraser Taylor (eds), *Development from Above or Below*, John Wiley, New York, pp. 39-72.

Taylor, D.R.F. and Mackenzie, F. (eds) (1992), *Development From Within: Survival in Rural Africa*, Routledge, London.

Time, 30, March 1998, 'African Rising', *Time*, Time International, Amsterdam.

Zvovashe, D. (1998), Personal communication, farmer in Bondolfi.

13 Subsistence and marginal lands – an Alaskan political and geographic issue

DONALD F. LYNCH AND ROGER W. PEARSON

Introduction

Alaska is a large peninsula on the Northwest corner of North America at approximately the same latitudes as Scandinavia but one third again as large. Alaska is bounded on the north by the Arctic Ocean, on the west by Russia and the Chukchi and Bering Seas and the North Pacific Ocean and on the east by the Yukon Territory and on the south and Southeast by British Columbia. Alaska's 1.5 million square kilometres of territory makes it seventeen percent of the United States. The seas around Alaska constitute one of greatest fisheries in the world, while the land supports wildlife including especially moose and caribou hunted for thousands of years by native peoples. Alaska became a state in 1959 with, however, most of its lands under the control of the Federal Government. Forty years and many laws later, the dominant landowner in Alaska remains the federal government which directly administers sixty percent of Alaska (see Figure 1). Alaska's population of about 625,000 people is concentrated in the greater Anchorage area, Fairbanks and Juneau, but approximately seventeen percent of its people are considered 'native' or indigenous, and they are the dominant people living on most of Alaska's land. However, increasingly native Alaskans are living in urban areas. Considerable fear has been expressed over the past two decades by native leaders that the urban white population may take over hunting and fishing on state and federal lands, depriving those who depend on wildlife resources of their livelihood, their subsistence.

The issue of who should control subsistence hunting and fishing on federally owned or claimed land and waterways in Alaska goes back to the

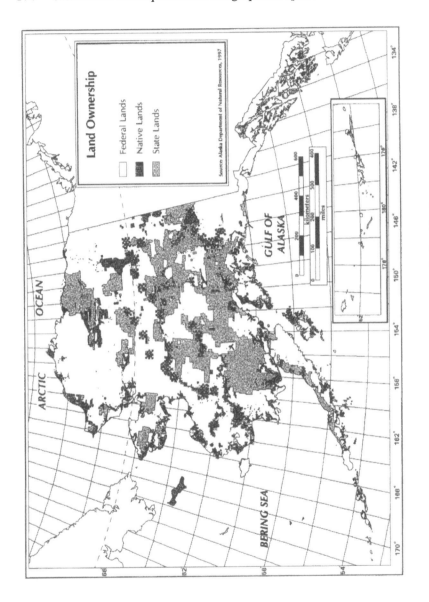

Figure 13.1   **Federal, state and native lands in Alaska**

Statehood Act of 1958 which granted statehood to Alaska in 1959. The major federal law governing Alaska's lands passed in 1980 reversed the Statehood Act, which many believe is a legal 'Compact' between the state and the federal government and, as such, should not be unilaterally changed. The Statehood Act granted fish and game management to the state over most land in Alaska. The 1980 law demanded a subsistence preference for rural residents, a provision which the Alaska Supreme Court held in 1989 violated the State Constitution. As a consequence, the federal government took over game management on federal lands in 1990 and fisheries management on sixty percent of Alaska's waterways on October 1, 1999. The state could have avoided both these actions by passing an amendment to the State Constitution approved by the Secretary of the Interior. The Alaska State Legislature has repeatedly failed to pass such an amendment including in a special September 1999 session.

The legal background of the subsistence issue

In 1980, the Congress of the United States of America passed a 448 page omnibus Act entitled the 'Alaska National Interest Lands Conservation Act' (ANILCA) subsequently ratified by President Jimmy Carter. The purpose of this Act was to settle federal land claims in Alaska and to set aside millions of acres of land for federal parks, wildlife refuges, conservation and national recreation areas, forests, wild and scenic rivers, wilderness preservation systems, and to settle various other land issues which had arisen due to previous federal laws. Title VIII of this Act concerned subsistence management and use and is the basis of the current and long standing dispute between the government of the State of Alaska and the Federal agencies managing federal lands in Alaska. Within Alaska itself, the issue of subsistence management and use has proven to be divisive. As the Alaska State Legislature and several governors have sought to find a compromise position amongst the various groups, the federal government taken over fish and game management in most of Alaska, a right given to Alaska under the Statehood Act. The Statehood Act was approved by the people of Alaska and is the law under which Alaska became a state in 1959.

Section 801 of ANILCA mandates a rural preference for natives and non-natives alike with a view towards continuing native 'physical, economic, traditional, and cultural existence' and non-native 'physical, economic, traditional, and social existence.' Congress argued that subsistence uses of resources 'on public and other lands in Alaska is threatened by the increasing population of Alaska, with resultant pressure on subsistence resources...' Subsistence itself is defined as 'the customary and traditional uses by rural Alaska residents of wild, renewable resources...', and such subsistence uses 'shall be accorded preference over the taking on such lands of fish and wildlife for other purposes...'

Under Section 802 (d) the Secretary of the Interior is directed not to implement federal management provided the State of Alaska 'enacts and implements laws of general applicability which are consistent with...' the provisions of ANILCA. Here is where the issue is raised. Initially the State passed laws and pursued policies which seemed to satisfy most of the federally mandated provisions, but these were struck down by Alaska's Supreme Court as violating the State's Constitution, particularly Article VIII, Section 3 which states: 'Wherever occurring in their natural state, fish, wildlife, and waters are reserved to the people for common use.' Article VIII, Section 4 further specifies: 'Fish, forests, wildlife, grasslands, and all other replenishable resources belonging to the State shall be utilized, developed, and maintained on the sustained yield principle, subject to preferences among beneficial uses.' In practice, the State had used Section 4 to maintain subsistence preferences without actually so stating.

The Alaskan issue, then, became one of amending the Constitution to bring it in line with the federal mandates under ANILCA. Article VIII of the State Constitution provides for two methods of amending the Constitution. One is a proposal by two thirds of the membership of both houses of the legislature, which is then to be placed on the ballot for the next general election. The other is a Constitutional Convention, which may be called by the Legislature, the recommendations of which are then subject to ratification by the people. Failing either of these two approaches, the Constitution also provides that every ten years the people will be asked at a general election whether or not a Constitutional Convention shall be held.

Under the Statehood Act, Alaska was given the power to manage fish and game over most lands in Alaska, a power which has been severely restricted by subsequent federal land withdrawals. The rationale, however, behind this provision in the Statehood Act was simply that federal fish and game management had proven to be a disaster prior to 1959 particularly in regulating the economically vital salmon fisheries of Alaska, which were in a catastrophic condition. Moreover, the most significant fish, salmon, migrate over very long distances, as do the major game animals, particularly caribou, bear and moose. Dividing management between multiple governmental entities it was thought would seriously inhibit wise and scientifically valid fish and game management. Further, the federal government had then a long history of inadequately funding fish and game management and saw state management as a means of reducing federal fiscal responsibilities in Alaska. State management has been associated with dramatic increases in the numbers of both fish and game in Alaska, in part, however, due to better regulation of salmon harvests in the Bering Sea under various international treaties. During 1997 and 1998. However, this pattern was reversed with significant and unexplained declines in some fish harvests particularly in the rich fisheries of Bristol Bay, which has lead to severe economic hardships in that region.

The political background

From March through early August 1998, and starting again in January, 1999, when the legislature reconvened, Alaskan newspapers were filled with articles regarding the subsistence issues. Various groups, whose memberships have not always been publicly revealed, undertook newspaper adds placing pressure on the legislature to pass a rural subsistence amendment to the Constitution, while native groups have held public demonstrations supported by the leaders of Anchorage's black clergy and other groups. The Governor and both Alaskan US Senators supported an amendment, but the legislature in its regular session and summer special sessions failed to obtain the necessary two-thirds majority.

The arguments against passing a rural, subsistence preference can be summarized as follows.

1) The federal government, by violating the Statehood Act of 1958, has performed an unconstitutional action, which should be appealed to the US Supreme Court. The state's legal advisors argue that such a case will fail, but that does not end the argument. The proponents of this argument feel that the Statehood Act is a solemn Compact or contract between the people of Alaska and the US Congress and can not be unilaterally altered.
2) Federal management of fish and game in Alaska, based on historical experience, will prove disastrous to fish and game stocks, an opinion strongly held by those who remember the conditions prior to statehood.
3) Alaskans should not be discriminated against by Zip Code, that is by place of residence, particularly since large numbers of residents in Anchorage, Fairbanks, Juneau and other cities do in fact rely on subsistence hunting

and fishing. What is not mentioned is that approximately half of the official native population of Alaska resides in Anchorage, Fairbanks, and Juneau.

4) State policies have traditionally favoured rural residents anyway, and the threat of Alaskan urban hunters decimating fish and game stocks to the detriment of natives is highly artificial, divisive and insulting.

5) Implementation of a rural subsistence preference will probably require some form of identification as to who is and who is not a 'rural resident,' as well as an effort to determine not only 'what traditional and customary use' has been, but also who has that entitlement. In short, a great new bureaucracy may be created, and the establishment of 'legal residence' will become complex and unfair, particularly to people who live sometimes in the cities, sometimes in the ' 'bush' or 'villages' or travel frequently, e.g. to work in the petroleum fields.

6) A constitutional amendment will apply equally to state and federal lands, and probably also to municipal and private land as well. Thus, while the federal mandate theoretically affects only federal land and water, the constitutional amendment would have affected all of Alaska.

7) A subsistence preference may lead to hunting year around or in areas which are suffering animal population declines, thus interfering (or possible eliminating) effective, scientifically based game management.

8) To compel a State to amend its Constitution in a manner mandated by the Secretary of Interior is an arbitrary and insulting abuse of political power.

9) Since the key subsistence fish migrate over long distances, imposing a subsistence preference on the federal rivers essentially mandates federal control of the state's coastal and riverine fisheries.

The arguments in favour of a subsistence amendment can be summarized as follows:

1) The federal law takes precedence over the state constitution and, therefore, to maintain state control over fish and game management the required constitutional amendment should be passed.

2) There is a real and present danger that urban and other sports hunters and commercial fishermen are now or may be in the future depriving Alaskans living in 'rural' areas of their basic foods which are obtained through hunting and fishing.

3) The refusal of the legislature to pass a subsistence amendment denies the people of Alaska the opportunity to determine whether or not they desire such an amendment.

4) The effort to oppose the amendment is basically anti-native and racist and is yet another effort by the Republican and Anchorage dominated legislature to oppose native rights.

5) A subsistence preference is a native right with a legal basis in the Constitution of the United States.

6) While not always explicitly stated, the legal reality is that a subsistence preference enshrined in the state constitution would apply to state as well as federal lands and probably also to municipal and private land, thus giving subsistence hunters preference throughout the entire state.

7) Non-natives are also at the risk of loosing their subsistence and are entitled to subsistence preference because they live in rural areas.

8) Store purchased foods, the substitute for foods obtained by hunting and fishing, are not only more expensive but far less nutritious, and so the issue is also one of personal health as well as the maintenance of native cultural values.

9) A federally mandated subsistence preference should create a significant number of jobs in fish and game management for Alaskan natives.

The first test case of the issue occurred in early August of 1998 in Northwest Alaska. The state, after consultation with local game boards, opened a subsistence season of nine months permitting hunters not using aircraft to harvest twenty nine Dall sheep and allowing eleven sports hunters selected at random to take full curl rams during a forty day season in the northern part of the region. This decision was overturned by the federal game board which denied the sports hunt, but permitted subsistence hunting by aircraft In effect, the federal government took over game management in Northwest Alaska.

Geographic issues

From the viewpoint of geography, one of the critical issues involved is what is meant by the term 'rural.' The word means living in the country as opposed to the city normally in an agricultural setting. This term manifestly does not apply in Alaska where agriculture, except for a few small areas, has not been and is not the basis for human settlement. The US Census bureau has used the concept that an urban place is one with a population of more than 2,500 people, in which case the major native settlements of Western Alaska would not qualify as rural, thus excluding Barrow, Kotzebue, Nome, and the native populations of Anchorage, Fairbanks, Juneau, Sitka and Ketchikan. However, the non-native populations of Chicken, Delta Junction, Glenallen, Copper Center, Cooper's (Cooper) Landing, and other road oriented, small places would clearly qualify. Thus, a subsistence preference, if this definition be accepted, is more likely to benefit non-natives than natives. Beyond that lies the question of how to certify where someone is a 'resident.' For example, many urban Alaskans maintain cabins in remote locations where they could claim residence, and similarly probably most urban Alaskan natives have relatives in and seasonally reside in communities, which would qualify as rural. The task of certifying who is or who is not a rural resident could be extremely complicated and introduce a whole new bureaucracy as well as tangled legal cases. Given these factors, the issue geographically can not and should not be viewed as a native versus non-native issue, and yet that is exactly the manner in which it is politically argued. [1]

Similarly, the definition of traditional and customary use is difficult to find. Does, for example, traditional and customary use include the sale of salmon roe and the use of thousands of salmon strictly for dog food, and does it include the use of expensive aircraft to go hunting hundreds of miles away from one's home? Does it include fur trapping as a source of income? These issues have already arisen, as has the question of deriving monetary income from wildlife. How much of the animal or the fish must one consume to qualify for subsistence, and how much can be sold? The question then arises as to not only what is 'traditional and customary,' but also who is entitled to subsistence hunting based on those criteria. Reliable data on past and current subsistence harvests of wild game have not been consistently collected, and what does exist relates to a small number of places. Defining, therefore, traditional use in a quantitative way is probably impossible. Once accomplished, however, the criteria could severely restrict future subsistence harvesting as population in rural communities continues to grow.

A critical geographic issue, however, relates to the migratory habits of the wildlife, which have no respect for political or game management boundaries. For example, a few years ago a herd of caribou from Mt. McKinley National Park over wintered on state, federal and municipal land in the Fairbanks areas. The question then becomes where does an animal have to be to qualify for subsistence hunting.

An excellent exposition of this issue comes from a Letter to the Editor of The Fairbanks Daily News Miner, June 9th, 1998. The writer is from Allakaket, a small native settlement on the Koyukuk River, a right bank tributary of the Yukon.

'I want to write a little bit about subsistence, and federal take-over. Around here at Allakaket, we have dual management for fish and game. The feds manage on their land and the state on state land. Neither agency is doing a good job, 'cause they don't come up with enough money for good management.'

'We only need one to do a good job.' This is kind of confusing too, especially for some of our elders who don't read or write. We have state land on one side of the river and the feds on the other side. State owns the gravel bars, up to the high water mark wherever that is. On the federal land we have a 'subsistence hunt'; a moose season that opens five days earlier than the state moose season. There is a little catch though. We have to shoot the moose on the federal land, which is on top of the bank of the river. We just hope the moose doesn't fall down the bank onto state-owned gravel bar. If this happens we will be in trouble. I am starting to lose my trust in the state Legislature, cause they have done nothing about the subsistence issue in the last 10 years. I don't think the state legislature will solve the subsistence issue in the near future, if they will, they would have done it already.

As I look to the south, I see dark clouds drifting north, I know the feds will be here soon.

Respectfully yours, *Pollock Simon Sr. Allakaket.*'

By way of explanation, moose are often hunted along the rivers from river boats, which makes transporting the dead animal much easier than tracking and then hauling the meat across country.

Another issue is who might be given the right to hunt animals, and specifically moose, in urban areas. The city of Anchorage has about 2,000 moose in residence, about 140 of whom are killed by automobiles annually. A similar problem exists in the Fairbanks area, and moose have become a major traffic hazard. Moreover, the urban environment, while free of wolves and other predators, does not always offer moose sufficient food, and so starvation is the frequent result. Failure to control such predators as wolves and bears due to national political pressure may be causing more and more animals to move into highly urbanized areas.[2] Does this mean, then, that subsistence hunters will have the right in a time of need to hunt and fish in urban areas?

Finally, will subsistence rights conflict with management strategies which seek to control the number of animals hunted by having open and closed seasons and limiting the number hunters may take? Will subsistence hunters now have the right to hunt and fish whenever and wherever they feel a necessity to do so? If this happens on a large scale, current fish and game management strategies could become null and void. The right to hunt out of season has been a periodic issue related to the native cultural desire to have fresh game for a potlatch marking the death of a relative.

The fisheries issue

Numerous fish are caught in Alaska for commercial, sport, and subsistence purposes including various species which inhabit only lakes and rivers. Possibly subsistence and sports users could be in conflict over the right to manage and utilize these species. The most critical fish are the various types of salmon, which are anadromous and move into the river mouths in a predictable pattern. Control of ocean fisheries is under federal supervision subject to international agreements, while heretofore management in the three mile Alaskan coastal zone and rivers has been under state supervision. Now sixty percent of the length of Alaska's rivers, i.e. those which transit federal property, are now under federal management. Mandating a subsistence priority along the federally controlled rivers is, however, meaningless, unless control is also exercised over how many fish are allowed upstream, known as escapement, which is under state control. The state has for most years successfully managed escapement to ensure renewal of fish stocks and has also with varying degrees of success been able to forecast future salmon runs. The role of federal management here could become quite significant since priority now will have to be given to subsistence users. This may also create a conflict in coastal fisheries, which are managed by having an open season, which is closed when the total catch reaches the established level. Subsistence fishermen may claim the right to fish even during closed seasons. Moreover, a certain portion of subsistence fish actually is sold commercially and this could be considered a 'customary and traditional' use. In short, the issue of when or what portion of a fishery is commercial and when it is for subsistence uses could become significant.

Subsistence and marginal lands

No matter which criteria one chooses to use, most of Alaska is marginal land: too high, too frozen, too wet, too cold, or too stormy for dense human settlement. The major commercial economic use for most of Alaska is mining for minerals and petroleum. Mining exists on a small scale basis in some areas, but major deposits are developed only where reserves are globally significant, e.g. the Fort Knox mine near Fairbanks, the Red Dog mine north of Kotzebue, as well as the Prudhoe Bay petroleum and natural gas deposits. In reality, therefore, the 'highest and best' use of most of Alaska is hunting and fishing.

To very influential national environmental groups, however, the highest and best use of most of Alaska's lands is to preserve wildlife and nature in their 'pristine' form, a concept which supports a significant and expanding, highly lucrative tourist industry and international sports hunting and fishing. As tourism continues to grow, a real conflict may develop between the demands of the tourist industry for a totally 'pristine' environment and the needs of local people to use the animals and other products of the land. This issue has already arisen about fur trapping which national groups are striving to prevent even though it is a most 'traditional and customary' use of wildlife and for many villagers and some urban dwellers a significant source of cash income. The November 1988 elections saw a referendum on the ballot to abolish the use of snares in trapping, a political effort funded by outside interests, but which failed at the ballot box. This undoubtedly is merely a first effort at attacking 'traditional and customary use' through political action.

The native settlements particularly in Western and Arctic Alaska enjoy substantial federal and state subsidies in the form of medical care, publicly financed housing, television, radio, subsidized electricity, and totally state funded education. The native corporations established by the 1971 Alaska Native Claims Settlement Act, which awarded natives 17.8 million hectares of land, provide significant financial support to native villages, while the Arctic Regional Corporation, based in Barrow, derives large revenues from taxation on the petroleum fields at Prudhoe Bay. One of the major consequences of these programs, particularly in public health, has been a dramatic increase in the number of natives, who constitute about seventeen percent of Alaska's resident population.

The increase in the numbers of natives has lead to expanded populations in such major native settlements as Barrow, Kotzebue and Nome, while smaller settlements continue to remain viable. As an example, Allakaket was completely rebuilt by the federal and state government, and with private donations from the people of Fairbanks, as a consequence of a flood a few years back. Without these subsidies, which by no means bring all modern amenities to native Alaska, the settlement pattern in its current form simply could not exist. The importation of food, fuel, machinery such as snowmobiles and aircraft and other commodities has become vital. At the same time, hunting, fishing and gathering are also mainstays of these settlements and indeed in most cases the only justification for their existence. Culturally, however, many native Alaskans prefer village to urban life, and those living in urban areas maintain their family ties with 'the bush.' Moreover, a large number of non-natives now reside in villages employed by government or in trade and services. This leads to an interesting question: should a highly paid white schoolteacher in a rural community be given a subsistence priority simply based on where he or she lives? And should the rural teacher have greater rights to subsistence than a lower paid teacher in a major urban environment?

For Alaska, marginal land does not mean land which is unused, but instead land which is subject to national, state and local political pressures for preferential uses. These disputes work their way through both the national and state political systems with some groups being politically more powerful than others, with local people appealing to the state or the federal government or national interest groups for support for their particular preferences.[3] In fisheries management, the key actors are not just the federal and Alaska state governments, but also the fishing interests of other countries, notably Canada, Japan, Korea, China and Russia, as well as the states of Washington and Oregon. With a federal take over of Alaska's fisheries, a new element may enter, that is a priority of preservation over conservation and subsistence, a thought which is beginning to enter the debate within Alaska. What then will be the subsistence priority vis-à-vis preservation and the fishing rights of other countries guaranteed by international agreements? Since under the US Constitution, Article VI, treaties take precedence over state constitutions and laws, an Alaskan subsistence amendment to the state constitution could not be enforced against foreign countries. This last issue has already been raised over migratory birds and subsistence whaling and may arise over management of caribou that migrate across the boundary between Alaska and Canada.

The Alaska lands issue historically goes back to 1867 when Alaska became part of the United States, and issues surrounding land use continue to be politically divisive. The subsistence issue has been debated for nineteen years, and significant corollary issues will keep it as a continuing political and geographical issue. The only ones who seem unconcerned are the moose, caribou, bears, wolves, deer, elk, birds and fish who do whatever they like regardless of the controversies surrounding human management of their welfare. In all probability, the real casualty is and will be scientifically valid fish and game management.

Notes

1. Members of the Kenaitze Indian Tribe based in the City of Kenai have argued that the Federal Subsistence Board should declare the entire Kenai Peninsula a subsistence region. At issue is fishing along the Kenai River, the most popular sports fishing stream in Alaska. Fairbanks Daily News Miner, May 10, 1999.
2. 'Game board votes to allow Anchorage moose hunt,' Fairbanks Daily News Miner, March 17, 1999.
3. 'Alaskan native organizations are raising funds to obtain political lobbying support from groups in the Lower Forty Eight to fight for subsistence rights or subsistence lobbying campaign,' Fairbanks Daily News Miner, March 18, 1999.

References

96th Congress, 2nd Session (1980), *Alaska National Interest Lands Conservation Act*, passed by the US Senate, November 12.

Fairbanks Daily News Miner, numerous articles from April through August 4, 1998.

Lantis, Dave, Donald Lynch, Roger Pearson (1981), 'Alaska: Land and Resource Issues', *Focus*, American Geographical Society, Vol. 31, No. 3.

Lieutenant Governor, State of Alaska (1996), *The Constitution of the State of Alaska*, Juneau, Alaska.

Roger W. Pearson and Marjorie Hermans, editors (1998), *Alaska in Maps*, University of Alaska Fairbanks, Alaska Department of Education, Alaska Geography Alliance.

14 Traditional water rights in northern New Mexico

OLEN PAUL MATTHEWS

New Mexico is a unique place for examining the sustainability of tradi-tional cultures. Hispano villagers and Pueblo Indians created a system of resource utilization that in some instances has sustained village life for over 600 years (Baxter, 1997; Meyer, 1996). These communities developed communal irrigation systems called acequias that allowed agricultural pro-duction in New Mexico's semi-arid climate (Nostrand, 1992; Meinig, 1971). In the process, a complex set of customs and laws were developed to allocate water between competing users and during shortages (Clark, 1987). When the United States gained possession of New Mexico, the property rights granted by the antecedent sovereigns in the region, Spain and Mexico, were recognized under the Treaty of Guadalupe Hidalgo and included the right to use water (DuMars et al., 1984). The territory and state of New Mexico also recognized these established water rights and the customary process of allocation between competing users that had been created. Although New Mexico claims to be a prior appropriation state in its water allocation, priority is only one of the factors considered under this traditional system. With increased pressure being put on limited water sup-plies, reallocation from traditional uses to urban uses seems inevitable.

When the Hispanos first explored New Mexico, they found village In-dians who were irrigating land (Carlson, 1990; Meyer, 1996). The irriga-tion tradition was well established at this time, and control over it was communal rather than being based on individual rights. The Indian lands were irrigated by ditches, and villages were located in those areas where water delivery was easy and protection from nomadic tribes was possible. At first the Hispanos located near Indian communities because they needed the resources these communities could provide as well as access to water. But the general colonization scheme was based on principles, which recog-nized the needs of both Indian, and Hispano communities to irrigate land (Baxter, 1997; Meyer, 1996).

The Hispano colonization process was state sponsored rather than individual as found in the United States (Pisani, 1992). Generally, a grant from the government was made to an individual or group of people before settlement occurred (Carlson, 1990). Some individual grants were made with the intention of organizing a community that would succeed to the rights contained in the grant. Those 'community' grants, whether originating in an individual or not, are the ones important to this study. These communities required irrigation. Before a 'community' grant was issued, the government required a determination to see if there was enough water to support the community (Meyer, 1996; Tyler, 1990). The grant was not to be issued if existing Pueblo Indian communities would be deprived of water. Although this principle of honouring pre-existing Pueblo Indian rights was often stated, in practice the results were not always clear cut, and disputes arose between Pueblo Indians and the Hispano population (Tyler, 1990). Additionally, the grant would not be made if interference with an existing Hispano community's water use would occur. Community grants were also made to Pueblo Indians under Spanish and Mexican law, but these grants were generally for existing communities. Pueblo Indians were given legal status, and their property rights were recognized and protected (DuMars *et al.*, 1984; Meyer, 1996).

Because New Mexico was far from existing food sources in Mexico, residents had to be self-sufficient. This made food production the first priority, and food production required irrigation. Many communities failed because water availability had been overestimated or changed over time. The second priority was protection from hostile nomadic Indians. To solve this problem colonists were required to settle as a group and organize their houses in a defensive square around an open plaza (Carlson, 1990). Some regions were occupied temporarily but abandoned and resettled at a later time when more peaceful circumstances prevailed. Community effort and co-ordination was required for survival. Although individuals received private rights to irrigable lands below the irrigation ditches (acequias), the acequia itself remained communal, as did grazing and timber lands beyond the land being irrigated. Community members who received water from the acequia also had an obligation to maintain it (Meyer, 1996). This combination of private and community property allowed cohesive, sustainable communities to form in areas where adequate water supplies and safe conditions existed (Baxter, 1997).

After a grant was issued, the first task in Hispano communities was ditch (acequia) construction, which often occurred even before house construction (Meyer, 1996). Typically, water would be diverted upstream from the village by using a small check dam made of rocks or brush. The water would feed into the acequia, which was a simple ditch built by hand with the tools available and was dependent on gravity flow (Clark, 1987). These primitive systems needed frequent repair. The acequias were generally not more than 3 to 4 kilometres long and carried water parallel to the river at the edge of the flood plain (Meyer, 1996). The fields between the acequia and the river could be flooded by gravity flow irrigation. Individual owner-ship of 'long lots' which ran between the acequia and the river facilitated irrigation (Carlson, 1990). By 1700, 62 acequias had been built with 102 more being added by 1800 (Clark, 1987). In 1974 about 1000 acequia asso-ciations continued in existence with the greatest number concentrated in Mora, Rio Arriba, Santa Fe, San Miguel, and Taos counties (Lavota, 1974). Although individual farms vary from one to 300 hectares, the typical range is 6 to 8 hectares. The number of users and the acreage under cultivation varies considerably from one acequia to the next.

As with any system of water use, conflicts arose which required the development of a system for resolving disputes between individuals and between communities. The resulting allocation policy developed under the umbrella of Spanish and Mexican law and was used to resolve disputes for both Hispano and Pueblo Indian communities. Because few lawyers were present in the region, the evolving 'law' included many practices that were in reality customs (Baxter, 1997; Clark, 1987; Meyer, 1996; Tyler, 1990). Today, litigation to resolve current disputes has made it important to de-termine the nature of the rights and customs established by Spain and Mex-ico (In re Application of Sleeper, 1988; New Mexico v. Aamodt, 1976;). As a result, a rich body of historic work has been developed to describe them. Historians, as expert witnesses, have been given an opportunity to study the evolution of water law and customs to an extent unseen else-where. A consensus is emerging on what the law was and how disputes were resolved historically.

Most local disputes, whether in Pueblo Indian or Hispanic communi-ties, were settled within the community (Tyler, 1990). Disputes between communities were more likely to go beyond the community boundary to a more formal legal process. Once a problem was brought forward for resolu-tion, a committee of 'experts' was usually formed to gather information and suggest possible solutions (Baxter, 1997). With this information in hand, the responsible official would make a decision binding on all – Pueblo Indian and Hispano. The principles used in resolving disputes have several aspects, but equity (a fair division of the resource) is the main one and includes principles of prior usage, intent, legal right, equality, rights of others, concern for the common good, need, and proportional sharing (Baxter, 1997; Tyler, 1990, DuMars *et al.*, 1984; Meyer, 1996). Much has been written about priority being the major factor in resolving disputes or sometimes the only factor (Clark, 1987; Lovato, 1974), but priority was only one element in these decisions. Certainly, if someone wanted a grant where there were prior users and insufficient water for a new community, the grant could be denied (DuMars *et al.*, 1984). The prior users would be protected. But this protection is really a matter of equity and not strictly priority. In other instances where insufficient water existed for all users, priority was considered but so were the needs of all the users. The common solution was to work out a method of sharing water that did not deprive anyone of basic needs. This is not the priority system that developed in the western states.

After New Mexico became part of the United States the pre-existing customary law became statutory and in the process was converted to something that looked more like the appropriation doctrine common in western states. The New Mexico territorial legislature was dominated by Hispanos during the early period, and they protected the status quo (Baxter, 1997). The best way to protect the communities was to give them a priority over other users. The appropriation doctrine does that. Within an aceqia community sharing is still done, and between Hispano communities informal sharing still occurs. But in disputes with non-Hispano users, priority of use is the dominant argument, because it generally puts the Hispano community in a better legal position.

When a more comprehensive water law was passed in 1905, the legislature exempted acequias from most of the requirements, and a different set of laws was developed for them (Clark, 1964; Lovato, 1974). One of the requirements is the eiection of a mayordomo (ditch boss) who sets up the rotation schedule for community users. Every April the mayordomo is required to consult with other mayordomos who get water from the same source in order to divide it between the different acequias. By state law, some counties have reduced the powers of the mayordomo and given it to a trio of acequia commissioners (Clark, 1987). The historic record demonstrates how past disputes were resolved (Baxter, 1997).

One of the major concerns today in Hispano communities is that individuals will sell water rights to someone outside the community. It is doubtful that the rights to sell water existed under Spanish or Mexican law, but this right is part of the appropriation doctrine. The communal right that existed under Spanish and Mexican law has been converted to a private property right. Potentially, such sales could destroy the community's ability to maintain the system and lead to the destruction of the community itself (Rivera, 1996). This possibility has become increasingly real as urban interests in Santa Fe and Albuquerque actively seek to purchase water rights.

The amount of water due a Pueblo Indian community is difficult to determine at this time because all their rights have not been quantified. At present, allocation between users within a Pueblo is occurring under the control of tribal governments and will continue in this manner. However, allocation between a Pueblo and an acequia or other user outside the Pueblo is not simple and may create long term problems in allocating water within Pueblos. During the Spanish and Mexican period, disputes between communities were resolved under Spanish or Mexican laws with the same 'law' applied to all. If one government body had jurisdiction today, the planning problems would not be as complicated. For example, if the state of New Mexico had jurisdiction over Pueblo Indians and New Mexico laws could be used to resolve conflicts, Indian Pueblos would be no different than the Hispano acequia communities.

But Indian water rights are a matter of federal law. Although Congress has given state courts authority to adjudicate Indian rights, the state courts must use federal law. For most Indian tribes, the reserved rights doctrine is the basis of the right. This right is based on 'practicably irrigable acreage', and was established by federal courts not Congress. However, most Pueblo lands were not set aside as reservations. In addition, Spain and Mexico recognized Pueblo rights, and they were validated with the Treaty of Guadelupe Hidalgo. Even though the rights have been recognized, the legal basis for the right and therefore the amount of water associated with them is unclear. A series of ongoing adjudications and law suits has advanced three possible legal theories for Pueblo Indian rights: aboriginal title based on usage before the colonial period, treaty rights based on the recognition of property rights created by Spain and Mexico, or something like reserved rights (DuMars *et al.*, 1984). No legal theory definitively controls how Pueblo Indian rights will be determined, but whatever the outcome, the allocation between communities will be affected as will the amount available within a community.

The status of Pueblo Indian water rights has been under litigation since 1965. At present the legal decisions are a mixed bag with one court saying they do not have reserved rights but may have something like them (New Mexico v. Aamodt, 1976). Another case says that the Pueblo rights are aboriginal rights (Reynolds v. Aamodt, 1985). Seven other cases are pending and will not be resolved until a final decision is made in the Aamodt case. Once a final decision is made and the extent of the Pueblo rights are known, the water may become available for reallocation. If reallocation occurs, the Pueblo communities may be seriously injured, because the resource that has sustained them for so long will be lost.

References

Baxter, John O. (1997), *Dividing New Mexico's Waters – 1700-1912*, University of New Mexico Press, Albuquerque.

Carlson, Alvar W. (1990), *The Spanish-American Homeland-Four in New Mexico's Rio Arriba*, John Hopkins University Press, Baltimore.

Clark, Ira G. (1987), *Water in New Mexico – A History of its Use and Management*, University of New Mexico Press, Albuquerque.

Clark, Robert E. (1964), *New Mexico Water Resources Law – A Survey of Legislation and Decisions*, Division of Government Research, Albuquerque.

DuMars, Charles T., O'Leary, Marilyn and Utton, Albert E. (1984), *Pueblo Indian Water Rights – Struggle for a Precious Resource*, University of Arizona Press, Tucson.

In Re Application of Sleeper (N.M. Ct. App. 1988), 760 P.2d 787.

Lovato, Phil (1974), *Las Acequias Del Norte*, Four Corners Regional Commission, Taos.

Meinig, Donald W. (1971), *Southwest – Three Peoples in Geographical Change*, Oxford University Press, London.

Meyer, Michael C. (1996), *Water in the Hispanic Southwest – A Social and Legal History 1550-1850*, University of Arizona Press, Tucson.

New Mexico v. Aamodt (10th Cir. 1976), 537 F.2d 1102 cert. denied 429, U.S. 1121.

Nostrand, Richard L. (1992), *The Hispano Homeland*, University of Oklahoma Press, Norman.

Pisani, Donald J. (1992), *To Reclaim a Divided West – Water, Law, and Public Policy 1848-1902*, University of New Mexico Press, Albuquerque.

Reynolds v. Aamodt (D.C.N.M. 1985), 618 F. Supp. 993.

Rivera, Jose (1996), *The Acequias of New Mexico and the Public Welfare*, Southwest Hispanic Research Institute, Albuquerque.

Rivera, Jose (1998), *Acequia Culture*, University of New Mexico Press, Albuquerque.

Tyler, Daniel (1990), *The Mythical Pueblo Rights Doctrine – Water Administration in Hispanic New Mexico*, Texas Western Press, El Paso.

PART 3
GLOBAL ECONOMIC DEVELOPMENT AND SUSTAINABILITY

15 Spatial shifts in production and consumption – marginality patterns in the new international division of labour

ASSEFA MEHRETU, BRUCE WM. PIGOZZI AND
LAWRENCE M. SOMMERS

Introduction

The spatial reorganization of post-Fordist production as influenced by the new international division of labour (NIDL) has produced a major geographical shift in employment opportunities at national and global scales. The first dynamic of this new geography of production is the growing spatial mismatch between the locations of production on the one hand and the locations of consumption and capital accumulation on the other (see also Dicken 1998, pp. 73-114; Barff 1995). Production has experienced a significant shift from old established industrial regions in the more developed countries (MDCs) to new ones, attracted by cheap labour pools and overall production costs, especially in less developed countries (LDCs). Consumption and capital accumulation have remained essentially anchored in the same prosperous regions of MDCs. This has made labour pools in rich industrial regions vulnerable to deindustrialization, down-sizing, and the resulting loss of employment in manufacturing and the downward levelling of wages. Concurrently, those regions also experienced tremendous growth in accumulation of capital by transnational corporations (TNCs) that reaped rewards from the restructured system of a flexible and globalized production. This contemporary dynamic in the growing spatial mismatch between production and consumption in the NIDL is contrasted with the high degree of spatial co-location of production, consumption and capital accumulation in MDCs that was characteristic of the old international division of labour

(OIDL). In other words, MDCs are beginning to experience some of the same mismatch between production and consumption that LDCs have experienced as primary staple exporters (Todaro 1994).

A second dynamic resulted in increased international convergence of production fuelled by foreign direct investment (FDI). This would have been favourable to international development. However, the resulting production shift benefited primarily the political and economic elite in both MDCs and LDCs whereas workers in both regions became highly vulnerable to degrees of exploitation and marginality. In the NIDL situation, market-led convergence in production may be taking place with flexibility, but wage and income disparities within nations and between nations have increased especially because of weakened unions in MDCs, and continuing elite hegemony in LDCs.

The complex nature of globalized production, consumption and accumulation than what can simply be explained by the new international division of labour must be recognized (Schoenberger 1988; Martinelli and Schoenberger 1991). The aim in this paper is to explore the nature of marginality that occurs in both MDCs and LDCs in the process of engaging the global economy within the rubric of the NIDL. For this reason, we begin by defining the meaning of *marginality* and suggest a typology of this concept paying particular attention to the nuances in patterns of exploitation brought about by the NIDL (see also Porter and Sheppard 1998, pp. 61-77; Leimgruber 1998). We close with a call for an attention on a new concept we term post-Fordist consumption, which suggests not only an increasing mismatch between the location of production and consumption but also a relative loss in aggregate demand by the working mid-to-lower income population.

International vulnerability to marginality

Marginality is defined here as a condition of socioeconomic distress experienced by people who are impoverished or lag behind an expected level of performance in economic well-being due to manifest vulnerabilities resulting from lack of competitive positions in free markets and or presence of oppressive hegemonic systems (see also Mehretu and Sommers 1994; Mehretu, Pigozzi and Sommers 1997; Friedmann 1988, p. 114; Gustafsson 1989, pp. 158-159; Gustafsson 1994, pp. 13-23; Blom 1998, pp. 164-175; Marcuse, 1996, pp. 204-211; McDowell, 1995). In the NIDL, factors of manifest vulnerability include the exposure of LDC labour pools to exploitative forays by TNCs to find the cheapest, most productive, and least organized (and less politically conscious) labour pools often dominated by the young and women (Dicken 1998, pp. 313-314). Important in this LDC vulnerability to marginalization is the collusion of TNCs with local elite that presides over a neo-colonial economy or a variation of crony clientelism and capitalism (Wild 1997, pp. 276-280; Galtung 1971).

In the case of MDCs, vulnerability of workers is a function of increasing rationalization and segmentation of industrial labour through a process of 'dualization' and 'bipolarization' (Castells, 1993, pp. 172-228). At present, a bipolar occupational structure would favour the employment of managers, professionals, engineers, and others that can be absorbed in high-technology-based industries. At the same time, those in companies that continue to depend on low skilled manufacturing jobs become vulnerable to downsizing and relocation. In North America, such bipolarization also has clear ethnic and gender implications in which majority males are favoured by this transformation whereas minorities and women become more vulnerable to marginality (see Castells, 1993, p. 179-184; Sassen 1996; Benko and Dunford, 1991, pp. 3-23).

Typology of marginality

In order to capture the nuances of interactive forces that produce marginality, a four-part typology is suggested. The two primary types are called *contingent* and *systemic* marginality. These two most important types give rise to a number of indirect marginality types, two of which will be featured in this paper; these are referred to as *collateral* and *leveraged* marginality.

Contingent marginality is marginalization of people/territories that spontaneously occurs as a function of free market mechanisms within a process of modernization and diffusion dynamics (Hirschman 1958; Myrdal 1957). Marginality within neo-classical variants is *contingent* because it results from *competitive* dynamics which have outcomes that maybe positive for some and negative for others. *Contingent marginality* results from *competitive inequality* and is considered to be endogenous to the neo-classical framework and is often addressed as 'accidental' or as a 'temporary and self-correcting aberration' in an otherwise equitable framework. It is also assumed that such 'aberrations' are expected to be taken care of by the 'self-adjusting' free market dynamics in MDC economies (see also Scott and Storper, 1992, pp. 3-24, Ernste and Meier, 1992, pp. 263-271; Marcuse, 1996, pp. 176-216; Wacquant, 1996, pp. 234-274).

Systemic marginality refers to uneven development caused by hegemonic forces of a political and economic system that produces *inequities* in the distribution of social, political and economic benefits. *Systemic marginality* may originate from forces internal or external to the state but in either case, its presence is determined by the degree of bipolar occupational and spatial structures that are present in the national economic system (Castells 1993, pp. 172-197). Unlike the neo-classical framework, *systemic marginality* does not lend itself to redressive reforms through "modes of regulation", as in the Keynesian welfare state, to mitigate the deleterious effects of unregulated enterprises (Painter 1996). In this case, marginality is a deliberate *construction* of an endogenous or exogenous system which intends to achieve specific desirable outcomes of political control (as in a dictatorship or pseudo-radical regime) or economic exploitation (as in neo-colonial regimes) (Dos Santos 1984; Riddell 1985; Mehretu 1989; Blaut 1993; Shannon 1996). This is perhaps the most important source of socio-economic marginality, especially in South countries as in the Sub-Saharan African region and Latin America. *Systemic marginality* is of particular significance in countries that have experienced pervasive inequity under colonial and/or neo-colonial regimes such as Zimbabwe and South Africa. Pseudo radical and totalitarian governments, common in South countries, often aggravate marginality by entering into lucrative compacts with foreign interests thereby reinforcing external influences of the national development process (see also Roseberg and Callaghy 1979; Mabogunje 1981; Rothchild and Chazan 1988; Young 1988; Friedmann 1988, pp. 108-144; Blaut 1993, pp. 17-30; Harris 1995). This is the case in the Congo in central Africa, a prosperous country with highly marginalized rural inhabitants whose wealth has been immensely squandered by a corrupt regime at the centre in collusion with extractive transnational corporate entities.

The first indirect form of marginality is *collateral marginality*. It is the systemic marginalization of people who, by virtue of their class or majority status, are not the intended targets of marginalization but because of their social or geographic proximity to vulnerable communities, suffer negative consequences. At the international scale, collateral marginality is not a major factor of inequality and inequity. It does happen in instances where core-country populations with no manifest vulnerability experience 'unintended' negative externality when they are present among people who have manifest vulnerability to contingent and systemic marginality. This may occur in the form of exposure to infectious diseases, random violence, terrorism, political instability, and unfriendly external linkages (development loans, FDI, etc.). The plight of missionaries, non-governmental and governmental volunteers in unstable environments may be characterized as collateral marginality.

A second indirect form of marginality is *leveraged marginality.* It is a much more significant in international interactions. It is a form of contingent or systemic marginalization of workers in one location whose bargaining position for improved social and economic benefits is weakened by the presence of impoverished workers in other locations experiencing primary contingent and systemic marginality. The root cause of *leveraged marginality* is the presence of core-periphery structures in global production relations with peripheral regions, mostly in LDCs, experiencing internally or externally induced economic as well as political hegemonies. *Leveraged marginalization* is a more recent phenomenon that has appeared with post-Fordist flexible production in which MDC workers are forced to compete with low-wage peripheral workers in poor regions or LDCs. An important factor in the new international division of labour is the enhanced ability of TNCs to threaten *leveraged marginality* on MDC workers by taking manufacturing jobs to off-shore locations where wages are low and controlled. This has of course been aided by major technological advances that have helped flexible production which has not only forced Fordist firms to vertical disintegration of organisation of production but also reduced the collective bargaining power of high-wage labour as down-sized firms began to relocate in areas with low-wage and less protected labour (Martinelli and Schoenberger 1991, pp. 117-124; see also Buck 1996; Gans 1996; Micheli 1996; Mingione 1996; Sassen 1996; Gottdiener 1994).

Spatial shifts and marginality

The nature of socioeconomic marginality is undergoing fundamental changes with shifts in post-Fordist and neo-Fordist patterns of production and consumption over the entire globe. New 'theatres of accumulation' have been formed with the rise in dominance of North America, the European Community and East and Southeast Asia as the 'world's triadic economic core.' Together these regions account for 87 percent of the world manufacturing value added and 80 percent of the exports. The current political, cultural and economic dominance of the west in reshaping the post-colonial and the post-cold-war world along core-periphery structures, resembles the geoeconomic and geopolitical implications of the Monroe

Doctrine of 1823 and the pole-to-pole pan-region idea of pre-war Germany (Taylor 1989, pp. 49-51). This has produced new dynamics in disparities of wealth and standards of living (Knox 1995, p. 232; Dicken 1998, pp. 60-62). There is also a belief that something fundamental has happened or is happening in world geography along geopolitical, geoeconomic, geosocial, and geoenvironmental dimensions (Taylor, Watts and Johnston, 1996, pp. 1-10; Dicken 1998, pp. 1-15).

One of the contributing factors to this fundamental change is the weakening of territorial integration of national communities for economic or social objectives under the onslaught of the TNC-led functional integration of the global market place (Friedman and Weaver 1980, pp. 163-188). This is sometimes discussed within the framework of the NIDL and 'new international financial system' that is orchestrated from global capitals with the framework of what has been called the 'Hymer' model (Dicken 1998, pp. 3; Porter and Sheppard 1998, pp. 61-77). The transformation from Fordist to post-Fordist spatial organisation of production and consumption has produced drastic changes in the world geography of marginality. OIDL marginality in MDCs was primarily contingent as differences in growth benefits were primarily a function of personal and/or spatial differences in competitive positions. In LDCs, OIDL marginality was mostly systemic because of the peripheral positions of most LDCs with most manifesting neo-colonial economic patterns and non-democratic hegemonic regimes. In the NIDL, marginality is taking a more complex process. Whereas LDCs continue to suffer from *primary systemic marginality*, MDCs are beginning to feel the pressure of *indirect systemic marginality* mediated by TNCs which are presiding over a globalized economy and a process of the internationalization of systemic marginality directly or indirectly (see also Porter and Sheppard 1998, pp. 61-64). This is what we refer to as *leveraged marginality*.

The post-Fordist period and its flexible· production design first and foremost began a process of dispersion of value added from traditional industrial hearths to peripheral locations nationally or to cheap labour pools in overseas areas. The working classes then became separated from a significant magnitude of the value added of what they consumed on account of outsourcing of components and/or direct imports of finished goods. In the US, the value added as percent of the gross domestic product, for agriculture, industry and manufacturing has declined. The value-added percentages for agriculture have declined from 3 percent in 1970 to 2 percent in 1998. For manufacturing, the respective declines were from 25 to 18 percent (World Bank 1993, 1999). At least in the short term, such reductions in value added means a higher level of competition that US labour faces with low-wage offshore producers. At the same time the share of value added for the service sector in the US rose from 63 percent in 1970 to 71 percent in 1998. But most of this change is due to growth in the high-paying service industry. The combination of high-level waged service sector and down-level waged manufacturing sector, led to the segmentation and bipolarization of the working classes facing dualized opportunities with most labourers under the pressure of leveraged marginality. This shift has also weakened trade union solidarity (see also Hayter and Barnes 1992).

Differential patterns in OIDL and NIDL

We are all familiar with the patterns of industrial production that have emerged from the OIDL as they have been summarized in classic 19th and 20th century location theory (Chapman and Walker 1987, pp. 16-52; Scott and Storper 1992, 3-13). Industrial production evolved as a vertically integrated and concentrated phenomenon in MDC industrial core regions. Most of what was produce at home was also consumed at home. Trade was primarily within national orbits or networks. Labour-management compact was characterized by co-operation with a relatively high share of the surplus value accruing to labour. Asset and capital accumulation was not only concentrated domestically but physical capital and financial capital were geographically nearly coincident. Over time the distribution of benefits reduced disparities and there was evidence of regional income convergence. Marginality was, at the macro scale, mostly contingent and endogenous. 'Use value' was essentially coincident with 'exchange value' especially for MDCs. Clearly marginality at the global scale was present with most peripheral regions experiencing largely systemic marginality (vestiges of the colonial and even mercantile), but much of it was also contingent due to a simple lack of participation in the industrial economy.

With NIDL many things have changed. Industrial production is flexible and footloose with spatial dispersion at the global scale being common (Chapman and Walker 1987, pp. 85-126; Asheim 1992). Consumption has become interdependent internationally and increasingly class stratified. Witness the increased marketing of such items as quality wines, BMWs and oriental rugs. In MDCs, labour-management conflicts have emerged and unions are weakened due to threats of outsourcing and the use of cheap foreign labour. In LDCs, labour-management hegemonies are rampant. Asset accumulation is flexible and no longer tied to financial accumulation.

TNCs, manufacturing as well as financial, span the globe with diverse ownership. This ownership is also 'class' and income stratified. Distribution of benefits is polarized within MDCs and LDCs. In MDCs there is a 'hollowing-out' of the middle class (Sassen 1996, pp. 64-82; Tosi 1996) especially with *leveraged marginality* while in LDCs the traditional polarity is simply intensified. 'Exchange value' predominates and exports are required in order to afford imports.

There are enormous socioeconomic consequences of NIDL. MDC economies have experienced considerable growth in their gross domestic product (GDP) whereas most LDCs have remained sluggish (World Bank 1999). Thus, at the global scale, the disparities will grow; the rich get richer and poor poorer. However, there will also be increasing disparities within both MDCs and LDCs. There will be greater income divergence. Trade will increase both in share and geographic scope with less 'use value' production. There will be a continued and increased internationalization in currency markets, ownership, and decision making.

Who loses? In the short run, not surprisingly, blue-collar labour pools in LDCs will continue to be subjected to systemic hegemonies by their own domestic elite acting in concert with TNCs. Already low-income levels will continue a downward slide with increased competition within and between LDC labour pools. Rural households in LDCs will continue to experience greater losses of environmental resources and the relocation of productive labour to urban activities rendering the population, especially women and children, more vulnerable to exploitation (Dicken 1992, p. 186, Buck 1996; Gans 1996; Micheli 1996).

Perhaps less obvious are losses in MDCs. Trade unions have come under increased pressure from leveraged marginality. This is a situation where MDC-based firms put pressure on workers and labour unions at home by threatening manufacturing relocation. Thus, a firm like Nike, Lee, Reebok or Arrow who have a history of production in places like Maine, Oregon, or North Carolina, will threaten the labour in those locations with the real possibility of moving that production to countries like Vietnam, Indonesia, Thailand, or Pakistan. This is a form of leveraged marginality because concessions are being leveraged in the MDCs on the basis of hegemonically impoverished labour pools in LDCs. Thus, primary systemic marginality experienced in LDCs is utilized to indirectly marginalize labour in MDCs.

Thus, the NIDL is contributing to a decline of the middle income groups in the rich countries with polarizing labour forces. These forces are impacting not just individual workers and entrepreneurs but also entire communities as tax bases erode. Those vulnerable to primary systemic marginalization within MDCs will be even worse off. Some ethnic and racial minorities, women, and perhaps older workers, will swell the ranks of those below poverty thresholds (see also Persson 1998).

A long term perspective is useful also. The OIDL is said to have developed in the context of Fordism. Fordist production has been described elsewhere and indeed, the phrase 'post-Fordist production' is nearly synonymous with globalization and flexible production (Benko and Dunford 1991, pp. 7-18). A focus on counterpoised ideas of *Fordist consumption* and *post-Fordist consumption* is appropriate. Henry Ford is credited with the entire assembly line model of manufacturing. But, he was very keenly aware that production was only part of what he needed to do. He was also famous for paying outrageously high wages in his day. He incurred the ire of his fellow industrialists for this. He did this not out of generosity. He paid high wages so his work force could buy the product he was making! Thus, it might be argued that *Fordist production* was matched, indeed, needed to be matched, by *Fordist consumption*; the workspace and the market space were highly correlated. Obviously, the NIDL is in the process of dismantling that identity. In the US we see the increased consumption of high-end products (BMW, Lexus, Rolex, etc.) while fewer workers can afford the new Ford or Chevrolet. With *leveraged marginality*, the NIDL may push this mismatch to a point where a *post-Fordist consumption* pattern will be acquired with potential relative reduction in income and a downward spiral of decreased aggregate consumption.

Conclusion

The post-Fordist geography of marginality has produced important international dimensions of inequitable development in both MDCs and LDCs. The first is the unregulated TNC role in carrying out undeterred flexibility with tremendous power to move both employment- and income-generating opportunities globally. This is made possible not only by technological advances for flexible specialization but also by the internationalization of systemic marginality. One dimension of this dynamic is the continuing systemic marginalization of LDC workers which is maintained by elite hegemony within LDCs. Another dimension is the enhanced role of TNCs to use leveraged hegemony on workers in MDCs.

Second is the loss of political clout of unions in MDCs to moderate the power of TNCs by forging mutually beneficial compacts in reshaping the NIDL. The power of the unions for effective collective bargaining has been diminished first by the ascendance of the political right, which has been hostile to organized labour in general, and second by new prerogatives of TNCs to refuse union demands with impunity. Loss of MDC union power also significantly reduced its role in advocating improved labour conditions in poor countries whose marginality would eventually affect their own welfare in MDCs due to immigration, low wages, and leveraged marginality.

Third is the weakening of regional and international organizations like the ILO, GATT, and WTO to put in safeguards against excessive exploitation of poor labour pools in LDCs without positive long-term impacts on the development of their economies. The non-existent power of such inter-governmental organizations (IGOs) on TNCs has rendered them ineffective in dealing with problems of unequal exchange and unfair labour practices, as TNCs are not responsive to the policy instruments of the IGOs. For this reason, income disparities between countries and within countries have worsened.

Fourth is the lack of national and international debate on the specific impacts of TNCs on society and space in both MDCs and LDCs with the aim of distinguishing responsible from deleterious flexibility in production and accumulation. The promise of post-Fordist flexible production was that it was finally unleashing the forces of trickle-down processes to move industrial production opportunities from the prosperous regions and countries to poorer regions and countries. What actually happened is that wealth became even more concentrated in rich regions and countries, and disparities between the poor and the rich in both LDCs and MDCs grew even wider.

Thus, the spatial consequences of marginality in the evolving worldwide shifts in production and consumption patterns are significant and rapidly growing. A major factor in this is the long-term implication of the post-Fordist spatial mismatch between production and consumption that threatens the growth of global production in absolute terms. Furthermore, the spatial mismatch between production and consumption has increased the vulnerability of workers, now in both MDCs and LDCs, to systemic and leveraged marginality. Wage disparities within nations and between nations are increasing. The reasons are contained in the globalization of production and consumption where labour markets are differentiated by wages and in power for collective bargaining. The situation has made it possible for TNCs, in their search for efficiency, markets and raw materials, to deepen *systemic marginality* in LDCs and introduce *leveraged marginality* in MDCs.

References

Asheim, Bjorn (1992), 'Flexible Specialisation, Industrial Districts and Small Firms: A Critical Appraisal', in H. Ernste, and V. Meier (eds.), *Regional Development and Contemporary Industrial Response: Extending Flexible Specialization*, Belhaven Press, London, pp. 45-63.

Barff, Richard (1995), 'Multinational Corporations and the New International Division of Labor', in R. J. Johnston, P. J. Taylor, and M. J. Watts (eds.), *Geographies of Global Change: Remapping the World in the Late Twentieth Century*, Blackwell, Oxford, pp. 50-62.

Benko, G. and M. Dunford (1991), 'Structural Change and the Spatial Organization of the Productive System: An Introduction', in G. Benko and M. Dunford (eds.), *Industrial Change and Regional Development*, Belhaven Press, London, pp. 3-23.

Blaut, J.M. (1993), *The Colonizer's Model of the World*, The Guilford Press, New York.

Blom, T. (1998), 'Microperipherality in Metropolitan Areas', in L. Andersson and T. Blom (eds.), *Sustainability and Development*, University of Karlstad, Karlstad, Sweden, pp. 164-175.

Buck, N. (1996), 'Social and Economic Change in Contemporary Britain: The Emergence of an Urban Underclass?', in E. Mingione (ed.), *Urban Poverty and the Underclass: A Reader*, Blackwell Publishers, Oxford, pp. 278-298.

Castells, M. (1993), *The Informational City*, Blackwell, Cambridge, MA.

Chapman, K. and Walker, D. (1987), *Industrial Location Theory*, Blackwell, New York.

Dicken, P. (1992), *Global Shift: The Internationalization of the Economic Activity*, Second Edition, The Guilford Press, New York.

Dicken, P. (1998), *Global Shift: Transforming the World Economy*, The Guilford Press, New York.

Dos Santos, T. (1984), 'The Structure of Dependence', *American Economic Review*, Vol. 40, No. 2, pp. 231-236.

Ernste, H. and Meier, V. (1992), *Regional Development and Contemporary Industrial Response: Extending Flexible Specialization*, Belhaven Press, London.

Friedmann, J. (1988), *Life Space and Economic Space*, Transaction Books, New Brunswick, NJ.

Friedmann, J. and C. Weaver (1980), *Territory and Function*, University of California Press, Berkeley, CA.

Galtung, J. (1971), A Structural Theory of Imperialism, *Journal of Peace Research* 21.

Gans, H.J. (1996), 'From "Underclass" to "Undercaste": Some Observations about the Future of the Post-Industrial Economy and its Major Victims', in E. Mingione (ed.), *Urban Poverty and the Underclass: A Reader*, Blackwell Publishers, Oxford, 141-152.

Gottdiener, M. (1994), *The Social Production of Urban Space*, University of Texas Press, Austin, TX.

Gustafsson, G. (1989), *Development in Marginal Areas*, University of Karlstad, Karlstad, Sweden.

Gustafsson, G. (1994), 'PIMA Research: Theoretical Framework', in U. Wiberg (ed.), *Marginal Areas in Developed Countries*, Umeå University, CERUM, Umeå, pp. 13-23.

Harris, Nigel (1995), *The New Untouchables: Immigration and the New World Worker*, Penguin Books, London.

Hayter, R. and Barnes, T.J. (1992), 'Labour Market Segmentation, Flexibility, and Recession: A British Colombia Case Study', *Environment and Planning C: Government and Policy* 10, pp. 333-353.

Hirschman, A.O. (1958), *The Strategy of Economic Development*, Yale University Press New Haven.

Knox, P.L. (1995), 'World Cities and the Organization of Global Space', in R.J. Johnston, P.J. Taylor and M.J. Watts (eds.), *Geographies of Global Change: Remapping the World in the Twentieth Century*, Blackwell Publishers, Oxford, pp. 232-247.

Leimgruber, W. (1998), 'Globalization, Deregulation, Marginalization: Where are We at the End of the Millennium', *Paper* presented at the IGU Commission on Dynamics of Marginal and Critical Regions, Coimbra, Portugal, August 24-29.

Mabogunje, A.L. (1981), *The Development Process: A Spatial Perspective*, Holmes & Meier Publishers, Inc., New York.

Marcuse, P. (1996), 'Space and Race in the Post-Fordist City: The Outcast Ghetto and Advanced Homelessness in the United States Today', in E. Mingione (ed.), *Urban Poverty and the Underclass: A Reader*, Blackwell Publishers, Oxford, pp. 176-216.

Martinelli, Flavia and Schoenberger, Erica (1991), 'Oligopoly is Alive and Well: Notes for a Broader Discussion of Flexible Accumulation', in G. Benko, and M. Dunford (eds.), *Structural Change and the Spatial Organization of the Productive System: An Introduction. in Industrial Change and Regional Development*, Belhaven Press, London, pp. 117-133.

McDowell, L. (1995), 'Understanding Diversity: The Problem of/for 'Theory'', in R.J. Johnston, P.J. Taylor and M.J. Watts (eds.), *Geographies of Global Change: Remapping the World in the Late Twentieth Century*, Blackwell, Oxford, pp. 280-294.

Mehretu, Assefa (1989), *Regional Disparity in Sub-Saharan Africa*, Boulder, CO: Westview Press.

Mehretu, Assefa and Sommers, Lawrence M. (1994), 'Patterns of Macrogeographic and Microgeographic Marginality in Michigan', *The Great Lakes Geographer*, Vol. 1, No.2.

Mehretu, Assefa, Pigozzi, Bruce Wm. and Sommers, Lawrence M. (1997) 'Issues of Urban Marginality in the Greater Detroit Area', in Gareth Jones and Arthur Morris (eds.), *Issues of Environmental, Economic and Social Stability in Development of Marginal Regions: Practices and Evaluation*, University of Glasgow and Strathclyde, Glasgow, Scotland, pp. 112-121.

Micheli, G. A. (1996), 'Downdrift: Provoking Agents and Symptom-Formation Factors in the Process of Impoverishment', in E. Mingione (ed.), *Urban Poverty and the Underclass: A Reader*, Blackwell Publishers, Oxford, pp. 41-63.

Mingione, Enzo (1996), 'Urban Poverty in Advanced Industrial World: Concepts, Analysis and Debates', in E. Mingione (ed.), *Urban Poverty and the Underclass: A Reader*, Blackwell Publishers, Oxford, pp. 3-40.

Myrdal, G. (1957), *Economic Theory of Underdeveloped Regions*, Duckworth, London.

Persson, L.O. (1998), 'Clusters of Marginal Microregions', in H. Jussila, W. Leimgruber and R. Majoral (eds.), *Perceptions of Marginality: Theoretical Issues and Regional Perceptions of Marginality in Geographical Space*, Ashgate Publishing Ltd., Aldershot, pp. 81-99.

Painter, J. (1996), 'The Regulatory State: The Corporate Welfare State and Beyond', in R.J. Johnson, P.J., Taylor and M.J. Watts (eds.), *Geographies of Global Change*, Blackwell Publishers, Oxford, pp. 127-143.

Porter, P.W. and Sheppard, E.S. (1998), *A World of Difference: Society, Nature, Development*, The Guildford Press, New York.

Riddell, B. (1985), 'Urban Bias in Underdevelopment', *Tijdschrift Voor Econ. en Soc. Geographie* 76, pp. 374-383.

Roseberg, C.G. and T. M. Callaghy (1979), *Socialism in Sub-Saharan Africa*, Institute of International Studies, Berkeley, CA.

Rothchild, D. and Chazan, N. (eds.) (1988), *The Precarious Balance: State and Society in Africa*, Westview Press, Boulder, CO.

Sassen, S. (1996), 'Service Employment Regimes and the New Inequality', in E. Mingione (ed.), *Urban Poverty and the Underclass: A Reader*, Blackwell Publishers, Oxford, pp. 64-82.

Schoenberger, E. (1988), 'Multinational Corporations and the New International Division of Labor: A Critical Appraisal', *International Regional Science Review* 11, pp. 105-120.

Scott, Allen, J. and Storper, Michael (1992), 'Regional Development Reconsidered' in Ernste, Huib and Verena Meier (eds.), *Regional Development and Contemporary Industrial Response: Extending Flexible Specialization*, Belhaven Press, London, pp. 3-24.

Shannon, T. R. (1996), *An Introduction to the World System Perspective*, Westview, Boulder, CO.

Taylor, P.J. (1989), *Political Geography*, Longman, New York.

Taylor, P.J., M.J. Watts and R.J. Johnston (1996), 'Global Change at the end of the Twentieth Century', in R.J. Johnston, P.J. Taylor and M.J. Watts (eds.), *Geographies of Global Change: Remapping the World in the Twentieth Century*, Blackwell Publishers, Oxford, pp. 1-10.

Todaro, M.P. (1994), *Economic Development*, Longman, New York.

Tosi, A. (1996), 'The Excluded and the Homeless: The Social Construction of the Fight Against Poverty in Europe', in E. Mingione (ed.), *Urban Poverty and the Underclass: A Reader*, Blackwell Publishers, Oxford, pp. 83-104.

Wacquant, J.D. (1996), 'Red Belt, Black Belt: Racial Division, Class Inequality and the State in the French Urban Periphery and the American Ghetto', in Mingione, E. (ed.), *Urban Poverty and the Underclass: A Reader*, Blackwell Publishers Inc., Oxford, pp. 234-274.

Wild, V. (1997), *Profit not for Profit's Sake: History and Business Culture of African En-terprise in Zimbabwe*, Baobab Books, Harare.

World Bank (1993, 1999), *World Development Report*, Oxford University Press, New York, NU.

Young, C. (1988), 'The African Colonial State and its Political Legacy', in D. Rothchild and N. Chazan (eds.), *The Precarious Balance: State and Society in Africa*, Westview Press, Boulder, CO.

16 Economic change and transportation deregulation in selected marginal and critical areas of the US Pacific Northwest[1]

STEVEN KALE

Introduction

The consequences of deregulation continue to trouble numerous analysts of social and economic change in marginal and critical areas. Deregulation is believed to place marginal and critical areas at a disadvantage because the private sector may choose to reduce or end services in areas where the government subsidized or required provision of services before deregulation laws were enacted. Rural areas are considered to be at higher risk than more populated areas due to smaller market size, low population density, and remoteness. Deregulation thus is believed to contribute to reduced well being in rural areas at the margins economically, geographically, and socially.

This paper explores the relationship between socioeconomic change and transportation deregulation in selected rural counties of the US Pacific Northwest, here defined as Idaho, Oregon, Washington, and western Montana. Economic change focuses primarily on changes in population and the number of employed persons from 1970 to 1995. Transportation deregulation includes reduced government regulation of companies providing for the movement of people and freight. Much of the transportation deregulation in the US occurred between the years 1970 and 1995, the same time period over which population and employment changes are examined.

The paper has several overall purposes. First, it explores the possibility that deregulation may have contributed to losses of population and employment in geographically marginal counties of the Pacific Northwest.

This occurs in part by identifying the extent of regional population and employment losses under the notion that deregulation potentially caused or contributed to socioeconomic decline. Second, for counties that declined in population and employment, the paper reviews various social and economic characteristics, including the transportation infrastructure. Third, the paper summarizes the provision of transportation services and where data are available, changes in service occurring after deregulation. Fourth, the paper assesses the information about socioeconomic change and deregulation in declining counties and speculates on whether deregulation contributed to decline.

Employment changes in marginal and critical counties of the Pacific Northwest

Marginal and critical areas of the Pacific Northwest are defined here as the 91 non-metropolitan (non-metro) counties outside the region's 23 metropolitan counties and 22 counties classified as non-metro adjacent according to the US Department of Agriculture (Butler 1990). Of the 91 counties, 37 are in Idaho, 22 are in Oregon, 15 are in Washington, and 17 are in western Montana. This regional definition is the same one used in one of the few geography textbooks for the region (Ashbaugh 1997). This textbook includes a chapter (Kale 1997) on the region's economy, including a description of changes in employment and income from the 1970s into the 1990s. Ohman (1999) also provides an analysis of social and economic change in the non-metro Northwest during the 1970s and 1980s.

According to data from the Bureau of Economic Analysis' Regional Economic Information System, the 91 non-metro counties in 1995 ranged from less than 1,000 to about 100,000 in population and from just over 500 to about 60,000 in employment. From 1970 to 1995, 10 counties lost population and seven lost employment. The following counties lost both population and employment: Clearwater and Shoshone in Idaho; Deer Lodge in western Montana; Gilliam, Sherman, and Wheeler in Oregon; and Garfield County in Washington (Figure 16.1). Among these seven counties, decreases in population ranged from 11 percent in Sherman County to 35 percent in Deer Lodge County (Table 16.1). Percentage losses in employment were similar to losses in population.

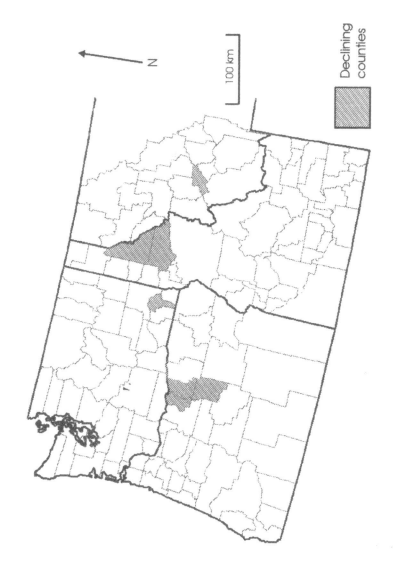

Figure 16.1 Counties declining in population and employment, 1970-1995

Table 16.1   Population and employment changes in declining counties, 1970-1995

County	Population		Employment	
	1995	Change, 1970-95	1995	Change, 1970-95
Clearwater, ID	9,202	-15.6	4,629	-16.0
Shoshone, ID	14,038	-29.3	6,042	-30.3
Deer Lodge, MT	10,200	-34.9	3,930	-37.2
Gilliam, OR	1,879	-16.6	1,196	-7.1
Sherman, OR	1,891	-11.2	1,137	-21.6
Wheeler, OR	1,564	-14.9	747	-15.0
Garfield, WA	2,289	-21.3	1,257	-26.2

Source: US Bureau of Economic Analysis, Regional Economic Information System

Clearwater and Shoshone Counties in Idaho have historically been de-pendent on the mining industry, primarily gold and silver from under-ground mines. A number of these mines have closed in recent years due to low mineral prices and a variety of environmental and other factors. Along with mine closures have come closures in support businesses, which has further contributed to declines. National forests cover much of these two counties, contributing to timber and wood products manufacturing, espe-cially in Clearwater County. Orofino, the county seat of Clearwater County, and several smaller towns are located on the Nez Perce Reserva-tion, which is mostly in neighbouring Lewis and Nez Perce Counties.

With nearly 19,600 residents in 1970 and just over 14,000 residents in 1995, Shoshone was the most populated county to lose residents and work-ers from 1970 to 1995. Clearwater County had nearly 10,800 residents in 1970 but by 1995, had dropped to a population of just over 9,200. Efforts to diversify economically have included the development of tourism, in-cluding water-based recreation in Clearwater County and the Silver Mountain ski resort in Shoshone County. Severe pollution problems from mining activities have hampered efforts to diversify in Shoshone County. Deer Lodge County in Montana experienced the greatest population and employment declines among the seven counties. From 1970 to 1995, population dropped from nearly 16,000 to just over 10,000; employment declined from nearly 5,850 to 3,930. Much of this decrease was due to the 1980 closure of the copper smelter in Anaconda (formerly called Coppero-polis). Efforts to diversify have helped slow the rate of decline but have not as yet resulted in a turnaround. Recent efforts include establishing the Smelter Stack State Park at the site of the former smelter, promoting numerous historic attractions associated with a former mining town, and developing Montana's first 'signature' golf course designed by Jack Nicholas. As in Shoshone County, severe pollution problems have hampered efforts to diversify.

Gilliam, Sherman, and Wheeler Counties in north central Oregon each have less than 2,000 residents and are the least populated counties in the state. They historically have been dependent on timber products and agriculture, primarily dryland wheat farming and ranching. As farms and ranches have increased in size, the number of people employed in agriculture and agriculturally related businesses has declined. Closure of timber processing operations in Wheeler County contributed to economic declines there. Formerly bringing in up to $1.6 million annually, timber fees from logging in national forests now total about $50,000 annually in Wheeler County (Senior 1999b). Efforts to diversify economically have included the development of solid and hazardous waste disposal facilities in Sherman County. Garbage from both Portland and Seattle are transferred to the landfill about 10 miles south of Arlington, Oregon. Since its opening in 1990, the landfill is estimated to have generated about 200 local jobs and accounts for nearly one-fourth of Arlington's city budget revenues (Oppenheimer 1999).

Garfield County in south-eastern Washington has just over 2,200 residents and, like Wheeler County in Oregon, is the least populated county in the state. Garfield too is dependent on dryland agriculture as well as a small amount of timber production in the southern part of the county near the Oregon border. The Snake River forms the northern border of the county, providing water-based recreational opportunities and a shallow-draft marine terminal for exporting grain. With Whitman County, Garfield County shares the Lower Granite Dam, one of four Snake River dams proposed for breaching to enhance salmon migration.

Transportation infrastructure and services

Interstate highways pass through Shoshone, Gilliam, Sherman, and Deer Lodge Counties. Only in Shoshone County does the interstate (I-90) pass through the largest communities and/or the county seat. Distances away from an interstate range from 8 miles for Anaconda, which is connected with a four-lane, divided highway to I-90, to nearly 40 miles between I-84 and Condon, Gilliam County's largest community and county seat.

Largest communities in the other three counties are considerably far-
ther away from an interstate, with Orofino in Clearwater County the farthest at about 150 miles. Major federally designated highways pass through
two of the three counties, with US 12 passing through both Clearwater and
Garfield Counties and U.S. 26 through the southern part of Wheeler
County but not through the county seat in Fossil, a town of about 500 residents.

Truck freight service is available to all of the counties. Several of the
smaller counties have very little freight to move except agricultural and
forest products, much of which is moved seasonally rather than year round.
Overnight parcel service is generally available to the larger communities
along major highways.

Russell's *National Motor Coach Guide* shows that intercity passenger
bus service is available along the interstate highways in Shoshone, Gilliam,
Sherman, and Deer Lodge Counties. Greyhound is the primary carrier providing service to these counties. A small local intercity carrier known as
the People Mover provides intercity bus transportation to residents of
Mitchell in Wheeler County. Neither Clearwater nor Garfield County has
intercity bus service. Each of the seven counties, however, has special
transportation providers serving senior citizens, the disabled, and other
travellers with special needs.

The Union Pacific's main east-west rail line in the Pacific Northwest
passes along the Columbia River in Gilliam and Wheeler Counties. Freight
rail service is provided to a grain elevator at Biggs and via a spur line to
the garbage disposal site about 15 miles south of the main line in Arlington. The Camas Prairie Railroad provides service to Orofino and other
communities in Clearwater County. Two short-line railroads--the Rarus
Railway Company and Montana Western Railway--provide service to Deer
Lodge County. There is no rail service to Shoshone, Wheeler, or Garfield
Counties.

Passenger rail service is not available in any of the counties. Until
May 1997, Amtrak ran the Pioneer Train on Union Pacific (UP) trackage
through Gilliam and Sherman Counties but didn't stop there. Efforts are
underway to restore Amtrak service between Portland and Boise, Idaho,
again on UP trackage through Gilliam and Sherman Counties (Senior
1999a). Currently, the nearest rail passenger service is provided by Amtrak
trains using the Burlington Northern and Santa Fe's rail line on the Washington side of the Columbia River north of Gilliam and Sherman Counties.
The nearest train station is about 25 miles from Biggs in Sherman County.
Anaconda is located the farthest from passenger rail service, over 200
miles. Amtrak also contracts with Greyhound and other carriers for inter-

line ticketing. Intercity bus passengers boarding in Gilliam or Sherman Counties can purchase Amtrak tickets through Greyhound agents, thereby facilitating boarding Amtrak trains in locations where rail passenger service is available.

While all of the counties have general aviation airports, none have regularly scheduled commercial service. Distances between commercial service airports and the largest community and/or county seat range from about 25 miles from Anaconda to Butte, Montana, to about 100 miles from Condon and Fossil to Pendleton and Redmond/Bend in Oregon.

Shallow draft (minimum depth of 14 feet) marine terminals are located along the Columbia and Snake Rivers in Oregon and Washington and as far east as Lewiston, Idaho, about 465 miles from the Columbia's mouth at the Pacific Ocean. Several companies operate grain terminals along the Columbia in Gilliam and Sherman Counties, including at the Port of Arlington in Gilliam County. Barge traffic also uses the Snake River along the northern border of Garfield County where the Port of Garfield is located. One of the Pacific Northwest's few dryland ports (Port of Montana) is located about 20 miles south-east of Anaconda. This is primarily a truck-rail terminal enjoying locational advantages of proximity to the Union Pacific Railroad, Montana Western Railway, I-90, and I-15.

Petroleum pipelines pass in or near several counties, but none provides direct service to communities in the seven counties. Natural gas transmission lines pass through Gilliam and Sherman Counties but do not provide service there. Kellogg and Anaconda, the largest communities in Shoshone and Deer Lodge Counties, have gas service but smaller communities do not. There is no natural gas service in Clearwater, Gilliam, Sherman, Wheeler, or Garfield Counties.

Deregulation and changes in transportation services

Many of the laws enabling economic deregulation in the U.S. were passed from about 1975 to 1985. These included the Railroad Revitalization and Regulatory Reform Act (1976), Air Cargo Deregulation Act (1977), Airline Deregulation Act (1978), Staggers Rail Act (1980), Motor Carrier Act (1980), Bus Regulatory Reform Act (1982), and Shipping Act (1984). More recent deregulation laws include the Federal Aviation Administration Act (1994) and the Trucking Industry Regulatory Reform Act (1994). In general, these laws allowed carriers to discontinue service for unprofitable areas or routes and loosened federal and/or state regulatory authority over the setting of rates.

The impacts of deregulation have been studied intensively. Many analysts have found that deregulation has resulted in more competition nationally and where competition has increased substantially, prices generally have dropped (Kale 1998). The research also has found that some types of service have declined and prices have increased, especially in smaller, remote communities with limited markets and few competitors. In some areas, services have been discontinued subsequent to deregulation. Examples of adverse impacts of deregulation on marginal and critical areas include abandoning rail branch lines and discontinuing passenger bus or air service.

For the seven declining counties in the Pacific Northwest, few changes in transportation services appear to have occurred as a result of deregulation. Although data are hard to find for the amount of services before deregulation, they appear to suggest that many of the smaller communities had few transportation options before as well as after deregulation. For example, none of the declining counties had commercial air service either before or after deregulation. Deregulation of the trucking industry may have even led to more competition in communities large enough to be of interest to motor carriers, many of which are small independent 'niche' operators that did not exist before deregulation.

Changes may have been greatest for intercity bus services and rail branch lines. Cullinane and Stokes (1998) have reported that bus deregulation in Great Britain generally has had negative effects on the provision of rural transportation. Similarly, in the US, the nation's largest bus company—Greyhound—substantially reduced the number of communities it served after the passage of legislation enabling intercity bus deregulation. According to Oster (1988), these reductions continued a decline in services that began before deregulation. The rate of decline, however, increased after deregulation. No research was found on changes in the daily frequency or timing of service. Fewer daily stops and changes from 'reasonable' timing (hour of the day) to 'unreasonable' timing of stops also would represent a deterioration of service.

In the seven declining counties, Greyhound continues to provide intercity bus services along the interstate highways and several of the more heavily travelled federal highways. The major exceptions are Clearwater and Garfield Counties which straddle US 12 in Washington and Idaho. Trailways, another major carrier, provides intercity service in Lewiston, Idaho, about 35 miles from the largest cities in Clearwater and Garfield Counties. Small intercity bus companies provide services in some areas abandoned by the large carriers. One community in Wheeler County was the only example identified where a small intercity bus company currently serves local residents.

Rail branch lines have been abandoned in several counties since 1970. Examples include the Union Pacific branch lines in Gilliam, Sherman, and Garfield Counties. A long-term trend of branch-line abandonment before deregulation suggests these lines may have been closed anyway. Railroads made few investments in lines generating little revenue. During grain harvest season, rail cars often were not available when needed, which contributed to even fewer car loadings and revenues. The major change associated with deregulation was that abandonment became easier than it would have been otherwise.

Other factors also have contributed to abandonment. Closure of a lumber mill in Wheeler County in the 1970s meant the end of all or most of the traffic for the Condon, Kinzua, and Southern Railway. This line linked to the UP's branch line in Gilliam County, so loss of the timber mill's traffic resulted in lower volumes and a less seasonally even flow of rail traffic over the UP's branch line. In Sherman County, abandonment of the UP's branch line occurred after a major flood washed out rail line trackage and bridges.

Deregulation of the rail freight industry contributed greatly to the start-up of small local carriers that have become known as short-line railroads, three of which operate in Clearwater County and Deer Lodge County. In some areas, short-lines have maintained or increased levels of service over what the major railroads provided. Service by short line companies is sometimes better than it was with the major rail companies which lost interest in short hauls and in maintaining branch lines. Improved service is difficult to provide unless enough revenues are generated to invest in trackage and equipment. Generating enough revenues for re-investment is difficult in areas where production in resource-based industries is declining or stagnant.

Discussion

Population and employment declines in the seven counties appear to be associated with a variety of factors other than deregulation. The declining counties in Oregon and Washington were the smallest in population in both states, suggesting small population size as a factor associated with decline. The four sparsely populated counties also are highly dependent on dryland agriculture, but this alone does not explain decline because other agriculturally dependent counties grew in population and/or employment.

The completion of major construction projects likely was a major contributing factor to decline in Garfield County and possibly in Gilliam and Sherman Counties. Construction of the Lower Granite Dam on the Snake River between Garfield and Whitman Counties was completed in 1975. Just downstream from Garfield County, construction of the Little Goose Dam was completed in 1970. After construction ended many workers and their families left the area which likely explains the majority of the decline. Similar effects may have influenced changes in Gilliam and Sherman Counties after construction of the John Day Dam on the Columbia River ended in 1968. Some of the effects may not have occurred until after the beginning (1970) of the study period for this study.

The decline in the mining/smelting industry appears to be the major contributing factor to population and employment losses in Shoshone and Deer Lake Counties and possibly in Clearwater County where declines in the timber industry may have been important too. In the last five to 10 years, each of these three counties appears to have recovered considerably from declines that occurred when major employers closed.

Little evidence was found to support the argument that deregulation contributed much to adverse effects on population or employment in the 91 most rural non-metropolitan counties in the Pacific Northwest. Over the 25-year period from 1970 to 1995, nearly 90 percent of the 91 counties grew in both population and employment. With so few counties declining, it does not appear that deregulation had serious net adverse effects.

This does not necessarily mean that individual businesses or families in some parts of the region did not experience ill effects. Closure of rail branch lines, for example, may have led to job losses for persons working in businesses directly dependent on rail service, but may have led to the creation of jobs in other businesses (e.g., trucking companies). Loss of intercity bus service may have reduced travel options for elderly, disabled, and other groups, but increases in special transportation providers may have lessened adverse impacts. To the extent adverse impacts occurred, they appear not to have had much effect on non-metro population or employment region-wide or in the seven counties with population and employment declines.

Just as numerous factors influence population and employment change, so to do many factors influence changes in prices and safety. Where freight rail service has ended, prices likely have increased for shipments carried by trucks rather than trains. At the same time, shippers may have switched to barge service where it is available, which likely would have lower rates than rail. Safety on roadways may have decreased slightly due to more trucks on the road after than before rail abandonment. Most of these changes likely would have occurred with or without deregulation in the railroad industry. To the extent that trucking deregulation has resulted in more trucking companies, some of which may operate with marginally adequate equipment, safety may have decreased. Increases in truck crash rates or number of crashes may suggest an adverse impact from trucking deregulation, but more investigation is needed to establish patterns, trends, and causation.

The low or negligible magnitude of adverse impacts from deregulation appears to hold for the seven counties declining in population and employment. The available data do not suggest much difference in the level of transportation services before deregulation than after deregulation. Communities that were poorly served after often were poorly served before.

Despite county population and employment declines, some communities continue to be served depending in large part on their location near major transportation routes. If they are near a major interstate highway, for example, they generally continue to have reasonable service given the size of their population and employment base. More extensive research might uncover additional examples of people and places that were negatively affected by deregulation, but there is little evidence to suggest that additional adverse impacts would be substantial.

Factors other than deregulation appear to have been substantial contributors to population and employment losses in the seven declining counties. The factors include the completion of major construction projects and closure or substantial downsizing of mining and related employers. During the 1990s, most of the counties appear to have stabilized from losses occurring in the 1970s or 1980s.

While the limited evidence presented in this paper suggests that deregulation is not a major contributing factor to population or economic decline in non-metro areas of the Pacific Northwest, it does not suggest that sustainability or globalization are unimportant to this region. Sustainability of natural resources such as salmon is an important issue for residents inside and outside the Pacific Northwest. Concerns include balancing the desire to sustain the salmon population with the desire to maintain local economic prosperity that depends in part on the availability of water for agriculture, barge transportation, and other types of usage.

Options to improve the number and rate of salmon migrating downstream are of immense interest to residents in the declining counties of Garfield, Clearwater, Gilliam, and Sherman. Efforts to improve the sustainability of salmon could lead to the breaching of dams on the Columbia and/or Snake Rivers. This, in turn, could lead to less irrigated agriculture, the reduction or closure of barge navigation, and higher costs of production and transportation in Oregon and Washington counties, including several experiencing relatively rapid rates of growth recently. 'Winners' from Snake River dam breachings, for example, could include salmon, commercial fisheries, outfitters, Native American tribes, railroads and trucking firms, road and rail construction companies, engineering firms, power companies not affected by loss of capacity, and port districts below the area of dam breachings (Barnett and Brinckman 1999). 'Losers' could include barge customers, barge operators, Snake River ports, irrigators, consumers paying higher electricity prices due to loss of hydro capacity, the Bonneville Power Administration, and river-dependent manufacturers.

Of the declining counties, adverse effects from Snake River dam breachings would potentially be greatest for Garfield County. A recent study for the Washington Legislature suggests that dam breachings could lead to transportation cost increases of up to 20 percent for Garfield County wheat growers (HDR Engineering 1999). This could result in even greater economic hardships for agriculturally dependent residents of the county. Adverse impacts might be reduced in part by increased travel-related expenditures by truckers of formerly barged goods, along with spending for highway improvements needed to accommodate increased truck traffic. Adverse impacts associated with dam breachings might also occur for farmers and timber producers across the border in Clearwater County, Idaho, along with decreased incomes from recreational activities due to drawdowns of the Dworshak Reservoir to help 'flush' young salmon downriver.

Similar and perhaps greater adverse economic impacts might be experienced in Gilliam and Sherman Counties if barge navigation is ended due to breachings or drawdowns for the John Day Dam and/or McNary Dam on the Columbia River. The US Army Corps of Engineers (1999) has already begun analyzing potential impacts from changes in the operation of the John Day Dam.

The argument could be made that sustainability is a far more critical issue than deregulation, at least in some of the marginal and critical areas of the Pacific Northwest. If efforts to improve sustainability result in higher costs of production and transportation, especially for the cheapest form of transportation--barging, then the benefits or negative aspects of deregulation are insignificant compared to the more substantial effects associated with other factors. Higher costs of production and transportation for agricultural and other resource based products in the region also would contribute to less competitiveness in the global economy, especially given the already low costs of production and/or the higher levels subsidization in some countries.

In closing, generalizations about the impacts of deregulation, sustainability, and globalization should be made very carefully based on facts and logic, not hyperbole. This paper makes a modest attempt at suggesting deregulation is not necessarily bad, strategies to achieve sustainability could have mixed and perhaps unintended adverse impacts, and deregulation and sustainability may interact in ways that require an open-minded evaluation. And last but not least, the effects of deregulation, efforts to improve sustainability, and globalization will vary spatially and over time, suggesting that taking a geographically broad and longer-range view should be considered equally if not more important than more narrowly conceived research ideas and hypotheses.

Note

1   The comments in this paper are the author's and not necessarily those of the Oregon Department of Transportation.

References

Ashbaugh, J., editor (1997), *The Pacific Northwest: Geographical Perspectives*, Kendall Hunt Publishing, Dubuque, Iowa.

Barnett, J. and Brinckman, J. (1999), 'Winners of a free-run river,' *The Sunday Oregonian*, July 25, 1999, p. A1.

Cullinane, S. and Stokes, G. (1998), *Rural Transport Policy*, Pergamon, Amsterdam.

HDR Engineering (1999), *Lower Snake River Drawdown Study*, Appendix B, Technical Memorandum No. 6, prepared for the Washington State Legislative Transportation Committee, Olympia, Washington.

Kale, S. (1997), 'The economy,' in Ashbaugh, J. (ed.) *The Pacific Northwest: Geographical Perspectives*, Kendall Hunt Publishing, Dubuque, Iowa.

Kale, S. (1998), 'Consequences of regulation and deregulation in marginal and critical economies of the U.S. Pacific Northwest,' Paper presented at the 1998 meeting of the International Geographical Union Commission on the Dynamics of Marginal and Critical Regions, Coimbra, Portugal, August 24-29, 1998.

Ohman, D. (1999), 'Restructuring and well-being in the non-metropolitan Pacific Northwest,' *Growth and Change*, vol. 30, no. 2, pp. 161-83.

Oppenheimer, L. (1999), 'Pride of place,' *The Sunday Oregonian*, August 29, pp. L1 and L8.

Oster, C. (1988), 'Is deregulation cutting small communities' transportation links?' *Rural Development Perspectives*, vol. 4, no. 3, pp. 13-16.

Russell's Guides (1999), *National Motor Coach Guide*, Cedar Rapids, Iowa.

Senior, J. (1999a), 'Portland-Boise rail plan on track,' *The Oregonian*, June 8, p. B2.

Senior, J. (1999b), 'Wheeler County: the Old West alive,' *The Oregonian*, June 10, p. C2.

US Army Corps of Engineers (1998), *Waterborne Commerce of the United States*, Washington, D.C.

US Army Corps of Engineers (1999), *John Day Lock and Dam Phase 1 Drawdown Study*, Portland, Oregon.

US Bureau of Economic Analysis (1998), Regional Economic Information System.

US Department of Transportation (1998), *Airport Activity Statistics of Certificated Route Carriers*, Washington, D.C.

17 Nearer the core – the fishing sector in one of Europe's marginal regions

XOSE SANTOS-SOLLA

Galicia in the context of Europe

There is little room for doubt about the geographical location of Galicia in the European continent: it is one of its 'land's ends', stuck in the Northwest of the Iberian Peninsula, in the Southwest of Europe. Without wanting to make easy recourse to physical determinism, this circumstance conditions many of the characteristic features of this land: the fact that it falls within the temperate climatic zone, with high rainfall, reflects on its landscapes and its culture; its physical isolation, aggravated by the presence of a mountain barrier hindering transport links with the East, has also caused greater contact through its southern borders with Portugal and through the North and West, across the ocean, with other nations from outside Iberian Peninsula. Many of the peculiarities of Galician culture were preserved thanks to the difficulties of contact with other Iberian peoples from inland Spain. However the difficulties brought about by this isolation and limited accessibility, in comparison to other European countries, were widely offset by the importance of maritime routes, above all as far as trade is concerned.

But this natural advantage was going to be neutralized by the political domination exerted by Castillians, which increases from the second half of the 15th century onward. It is true that the role of the Spanish sea front is reinforced during that period, but for the exclusive benefit of Andalusian ports. On the other hand, constant European wars are notably detrimental to traditional Galician trade. There is even a history of labour recruitment for work aboard ships at the service of the monarchy. Perhaps the most remarkable deed during this period is the creation of an increasingly centralized and geographically centred peninsular state. Peripheral areas become even more peripheral, particularly Galicia, with hardly any privileges granted on its ports, very weak commerce and no political capacity for ruling its own destiny.

The onset of the contemporary era brings about a worsening of the factors outlined above. Both Spain and Portugal appear consolidated as two very centralized states with their decision-making headquarters a very long way from the North, particularly from the Northwest, a dead end. Moreover, sea routes start being replaced at this point by terrestrial routes with the advent of the railways. Interest in the sea remains for long-distance exchanges but, with the loss of the colonies and the economy being highly protectionist, overseas trade is meagre. On a different front, the railway linking Galicia with central Spain did not arrive until the final quarter of the 19th century. Furthermore, the train was too slow and expensive to be profitable. However, profitability did not matter much given the poor level of industrialization of Galicia.

To sum up, Galicia today, leaving aside the internal differences, appears as one of the most backward territories of the European Union and, according to recent reports, as one of the areas with the worst socioeconomic prospects. The EU, in its search for cohesion, is developing programmes designed to reduce the differences separating richer and poorer regions. The inclusion of Galicia within the group of 'Objective 1' regions responds to that aim. Unfortunately, aid in the form of development programmes, apart from the fact that it might not always be used adequately, does not make up for this peripheral situation. We would even go as far as to suggest that European policy, despite its declared intentions, clearly benefits the richer regions of central Europe. The cohesion funds Spain fought for so eagerly alleging its condition as a poorer member of the Union, also benefit the infrastructure projects of the richer parts of Spain, which are growing, thus increasing the distance between one part of the country nearly at the same level of development as the rest of Europe and another, the peripheral one, which is even further away from reaching it. Galicia finds itself included in this last category.

Fishing within the context of Europe

In spite of the domination over the seas long present in European tradition, with powers as important historically as the United Kingdom, the Netherlands or Portugal, fishing has always been a very ancillary activity for most of these countries. France, Great Britain and Germany, the ruling characters on the Western European stage, have developed a very strong agricultural and farming economy. The sea was of interest merely for its function of providing commercial routes. Fishing was thus restricted to some coastal regions and peoples on the Atlantic front; one could even speak of hunting rather than fishing, given the importance of whaling and the industrial applications derived from this activity.

The scarce relevance of fishing in most of Europe is reflected in aspects of cultural behaviour as significant as gastronomy. The data concerning human consumption of fish (Varela *et al.*, 1997) are very revealing: Most European countries do not surpass 20 kilograms per person per year, and in some places like Germany or the Netherlands, levels are at less than 10 kilograms. Obviously, there countries with notably higher levels of consumption: Portugal, for instance, with 60 kilograms per capita, the same level as Norway and, above all, Iceland. The latter reached a consumption of 100 kilograms in the sixties. It is true that in recent years a substantial increase in fish consumption has been observed, owing to an improvement in the diet of Europeans. Nevertheless, this is not the general rule. As early as 1972, Coull pointed out that in the most industrialized countries, consumption preferences moved towards meat and eggs, particularly as per capita income increased. As an example he chose the United Kingdom, where fish consumption had been reduced by half since the 1920s. Most recent data also highlights this standstill, just as in Belgium, Germany and the Netherlands.

In any case, it is interesting to underscore the notion that fish consumption in Europe is relatively low, although it is on the increase in most countries. Of course, if we focus on individual regions, the situation changes notably.

Another interesting detail referring to the use of fish in the diet of Europeans (Varela *et al.*, 1997), measured in terms of calorie content, is very low, in many cases lower than 1%; in terms of protein consumption, Fish accounts for more than 10% in countries like Denmark or Portugal. This reflects, once more the scarce influence that fish has in general in European culture, with one or two notable exceptions. In any case, compared to other developed countries such as the United States and Australia, consumption is much higher.

The reasons accounting for this fact are complex: perhaps we could refer, as is the case with Spain, to a tendency to establish links with the interior of the continent. The greater security in the face of invasions coming from the sea, the wide agricultural and farming possibilities offered and the centralizing endeavours states had on their territories, can be used as explanations. On the other hand, the perishable nature of fish made it difficult for it to be transported to more densely populated inland areas. The sea remained therefore as provider of commercial routes and most ports were commercial rather than anything else. Only in countries with a strong Catholic tradition, such as Spain or Portugal, imposing a very strict calendar on fasting, fish consumption was somewhat higher.

It is rather difficult to quote statistic data about fishing, at least concerning figures going beyond commerce or the unloading of catches. All too frequently, information is very vague, it appears mixed up with agricultural data or else does not exist at all. A publication by Eurostat (European Union Bureau for Statistics) literally read 'fishing activities are particularly important in numerous less developed regions where alternative employment does not abound' (Europa en Cifras, 1992, p. 166). From this statement we can draw the conclusion that fishing is an activity relegated to marginal regions, which is maintained simply because of the lack of other employment opportunities. Also published by Eurostat, we find the yearbook of fishing statistics (1994) containing rather meagre information relating to catches for each fishing region, the species caught, and information about the fleet size and levels of outside trade.

This scarcity of figures clearly reflects how little importance fishing has in Europe. This is in sharp contrast to the relevant role the European Union has to play in the organisation of fishing and the signing of international agreements (González and Macau, 1996). Indeed, the majority of fishing grounds under European jurisdiction are the subject of negotiation between different countries with interests in them. This circumstance gives rise to tensions which in recent years have materialized in conflicts between states such as Ireland, Great Britain, France, Spain and Portugal. Besides, there are numerous European ships fishing in international waters or in other countries' areas of exclusive economic interest. In either of these two cases, but mostly in the second one, it is necessary to carry out negotiations that allow all concerned to reach beneficial agreements.

Generally speaking, there are several kinds of bilateral agreements between the European Union and other countries concerning fishing matters, resulting from very different approaches. In some cases, access to fish quotas is part of a global economic negotiation; in other cases, there is straightforward reciprocity: more or less balanced quota exchanges. With poorer countries, above all from Africa, the relationship tends to be purely monetary, that is, through the payment of money in exchange for access to fishing resources. Obviously, not everything is so linear in the sense that for each particular case slight novelties can be introduced which complicate the signing of the agreement. The last kind of agreement, which is regarded as the most complete and promising (González and Macau, 1996) is one which implies aid for development. They are second generation agreements that involve, for instance, the establishment of mixed-nationality companies or investments in other countries.

As can be observed, the rule the European Union plays in fishing matters is fundamental. Let us not forget either that many international fisheries are regulated and that the Union legally represents its members. This is the case, for example, with the NAFO waters near Newfoundland. As we said before, because of conflicting interests, there is continuous tension between member states of the Union: accusations of the use of illegal nets, of surpassing the allotted quotas, and of not complying by the fleet reduction programmes are very common and conceal obvious economic and strategic interests related not so much to fish consumption but to commercial exchanges and the creation of employment. Moreover, these kinds of conflict are used to arouse false nationalist passions in some states, in the face of the purported looting of their natural resources by foreign ships (Santos, in print A).

In spite of everything said above, the context of fishing in the European Union constitutes a very restricted area and its importance, we insist, is very limited. Recent rumours (fortunately unfounded) relating to the disappearance of the Direction-General XIV of the European Commission, which deals with fishing affairs, that would merge with the Direction-General for Agriculture, are only one more example of the scarce transcendence fishing is assigned, as well as the little political resolve detected when it comes to solving conflicts which should initially not be excessively complicated.

Fishing within the Galician context

Within the European context, Spain does not differ much. No one is a stranger to the fact that Spain is a Mediterranean State and that we are talking about a commercial sea where fishing has scarcely any relevance. There is no Ministry of Fishing. On the other hand, notorious conflicts such as the one about Turbot catches or the use of huge dredging nets known as *volantas*, or those arising from time to time with Canada, Morocco or Argentina are minimized within the context of state policy. Even Spain's accession to the Union in 1986 followed the waiving of recognized historical fishing rights and the acceptance of an excessively long transition period which involved that, at the end of it, Spanish ships were denied access to certain European Union fisheries, very specially the rich and well known Irish Box (Xunta de Galicia, 1993). Within this environment, the position of Galicia is very different. As we said at the beginning, Galicia is a land which has always been linked to the sea and fishing is an inseparable part both of its economy and of the lifestyle of many of its coastal villages. In this way, this activity turns into a fundamental pillar supporting the survival of an important proportion of men and women. We said before that Galicia is an Atlantic country embedded within a Mediterranean state. This statement, valid for fishing but also for other areas of our productive and socioeconomic system in general, creates significant and serious confrontations between the different territories making up the Spanish State: Atlantic agriculture versus Mediterranean agriculture, fishing harbours versus marinas and commercial harbours, fishing seas versus seas for tourists (Santos, in print B). These contradictions are solved frequently on the basis of the economic and political weight of the different Autonomous Regions and of the strategies of the state which are, by the way, clearly Mediterranean-oriented. As a significant detail to be borne in mind, we could mention that Galicia accounts for 7% of the Spanish population and that as far as its per capita income it is among the last positions of all Spanish Autonomous Regions.

Fishing is, without doubt, one of the most disfavoured sectors. In order to have a better appreciation of its context in Galicia, we can offer some data indicating the importance it has for Galician economy. For Spain as a whole, fishing contributes merely 0.4% of the GNP, and 3% in Galicia; over 40% of Spanish crewmen are Galician and only the Galician fleet measured in Gross Registered Tonnage exceeds that of any European Union Country. These are some of the parameters indicating the relevance that fishing has for Galicia. We are not taking into consideration other circumstances complex to measure related to the globalization of the economy. For example, we know that a substantial number of crewmen in ships registered in other Autonomous Regions, such as Andalucia, the Basque Country and the Canary Islands are Galician, as are the ships and the capital. We also know of the existence of Galician ships fishing under flags of convenience or through the formula of mixed-nationality companies or simply through direct investment in those countries. All that contributes very notably towards fishing having great weighting on the economy and employment, even without considering the wealth it indirectly produces in the service and industry sectors. To sum up, we are talking about an activity of strategic importance for Galicia from any point of view (its contribution to Galician GNP is estimated at 10%). Getting down to specific details, certain coastal towns and villages base their economy, totally or partially, on fishing and its derived industries.

This economic dependence on fishing, also finds reflection in another kind of dependence: a political one. It must be considered that the narrowness of the Galician seaboard has long pushed Galician sailors to earn a living in distant waters. Hence, most products of Galician fishing are obtained from fishing grounds which are either international or under another country's economic jurisdiction. In the latter case, it is the responsibility of the European Union to reach agreements. Whereas Galicia has no representation as such in this institution, it is Spain that takes part in those decision forums. This fact must be viewed within the status quo of the scarce importance fishing has both in Europe and in Spain, which significantly conditions negotiations. How can it be brought home to European Union Authorities that such a central topic for a peripheral territory must be given adequate attention, when the prevailing inertia dictates that it really has no transcendence? For instance, how can we get the European Union or the Spanish Government to really become concerned with Argentina's volubility regarding mixed-nationality companies (a large part of which are Galician) or with the constant difficulties created by Morocco or Canada's decisions regarding international waters?

The proposed solutions to this problem are multiple. One consists of the creation of a lobbying group, both in Madrid and in Brussels, which sees to defending the interests of the sector. On a political front, for example, we have the Fishing Direction-General of the Spanish Ministry of Agriculture in Galician hands, in the same way as the Fishing Commission of the European Parliament is presided by a Galician woman. Apart from these two cases (another attempt to place a Galician man at the helm of the Direction-General XVI failed), there is a fundamental lobby which has achieved big successes, such as the change of some discriminatory articles for the Galician fleet in the British Merchant Shipping Act, including the payment of compensation for the damage caused by the inactivity of the fleet. Another kind of measures are those to do with the signing of private agreements prompted by companies very often with the support of the Galician Government. These measures have given rise to interesting covenants favourable to all parts concerned since access to fisheries is usually accompanied by actions for development, such as investments in other countries, thus creating wealth and employment.

In a nutshell, multiple means have been employed in an attempt to solve the fishing difficulties encountered by Galicia, as we have seen. Unhappily, the desired results are not always reached and fishing continues to be a very marginal matter within the European Union and the context of Spain itself, even in spite of the fact that the consumption of fish is experimenting substantial growth. Now we are going to focus on the issue of the Galician fishing tradition with the aim of analyzing the context of globality produced within the scope of enclosed local communities, such as fishing communities are.

There are numberless historical testimonies about the consumption of sea products, proofs coming from as early as prehistoric times (Calo, 1996). From a geographical perspective, the location of numerous population centres on the coast, is a sign of inevitable contact with the sea. As opposed to the Mediterranean Sea, commerce this side of Iberia, on the Atlantic coast, was far less important, and restricted to some towns and villages with special privileges. Most of the villages located by the sea found their *raison d'être* in the food sources afforded them by the Ocean. These small settlements (most of them with only a few hundred inhabitants) used to be very isolated from the outside world. The abrupt relief of Galicia, above all that of the coast, explained this isolation. In this way, secluded communities became consolidated where the sea was the main source of the economy.

With the passage of time, now nearing the 20th century, in these villages there arose two features that were to define them perfectly. On the one hand, the over-exploitation of fisheries and improved technological methods were to prompt and facilitate the ships faring to distant waters. On the other hand, the isolation of many of these settlements is going to be reinforced: they are not reached by train and they are served by really precarious transport links. As the only salient novelty, the canning industry now allows a little trade, but this is going to develop, mostly, in the best-linked coastal areas.

These characteristics became more apparent as the century advanced. We can make special mention of the 1960s, when technology allowed the advent of large freezer boats that journey away on fishing campaigns for several months sailing across the seven seas. Besides, the development of an oil-based economy and therefore of the merchant navy makes many seamen choose this much better paid option. As a consequence, not only are they going to become acquainted with new countries, but they are also going to share space and experiences with seamen from many other nations. Norway, Japan, South Africa and Canada are some of the most common destinations of their voyages. Meanwhile, on land, we observe the growth and consolidation of a North-South axis running across the southern coast, but leaving aside a great part of Galician coastal areas, which are still isolated and badly linked with the country's main growth areas. All these facts contribute to reaffirm a historical tendency which, nevertheless, belongs in the 20th century: that is, the 'global versus local' duality. Seamen show a deep knowledge of the world thanks to their personal experiences but, on the other hand, they are inserted in enclosed communities where social networks become fundamentally important.

In our paper (Santos, in print B), we could confirm that the working life of Galician seamen starts very early. Until quite recent times (the 1970s), it was common for them to learn their skills with the aid of their grandfathers who took them to sea on small boats, always near the coast. Frequently, their age did not surpass 10 or 11 years. Then they were usually taken on board larger boats, fishing not too far off the coast. Their next step, once trained in the profession, was to work on deep-sea fishing vessels a long way off the shore, or also in the merchant navy. Times of bonanza (good salaries and fleet expansion) allowed some of them to buy a small boat or set up a little business, generally a bar. In any case, in the latter years of their active lives, they usually went back to coastal fishing, falling back on artisan methods after retirement and passing their skills down to their grandsons.

This situation has now substantially changed. Although the learning of fishing skills may still maintain some traditional features, official training schools play an increasingly important role. In any case, regulations on the minimum working age prevent youngsters from having access to fishing until completing their compulsory education. Another fact to be taken into account is that there is hardly any offer of employment for Europeans in the merchant navy, but mostly for Asians. Besides, one of the main problems affecting fishing today in Galicia is the scarce offer of manpower for such a dangerous job[1], which is poorly paid[2] and involves too hard an emotional and physical effort. In this last respect, we must take into account the age of retirement which, as for miners, is set at 55 years. On the other hand, long absences provoke very unstable family situations (Zulaika, 1993); at the same time, life together on the boats (where the physical space is very reduced) for months on end, provokes numerous tensions. At present, the profession of fisherman appears reserved for those who are incapable of finding an alternative occupation, or who do not want to take their education further, or who want to earn what is a relatively good salary considering their age and qualifications.

Today, one could talk about a general crisis affecting a large section of fishing communities, at least the most isolated ones. The disappearance of fishing, difficulties in accessing international fisheries, or the rejection on the part of young people of the profession of fisherman itself are some of the factors affecting these settlements. On the other hand, the absence of economic alternatives, aggravated by precarious transport links, pushes the population towards emigration. This affects men, but most clearly women. Females, with practically no employment alternatives and showing a growing tendency to avoid falling into a relationship with a fisherman (because of the implications of hardship this involves for them) have very high rates of emigration.[3] Women have traditionally played a very important role in fishermen's settlements (Pinheiro, 1997). Beyond the usual domestic chores, they carried out essential tasks such as the weaving of the nets, the selling of fish, which represented an extra income, the collection of molluscs and part time agriculture. Besides, the men's prolonged absences forced them into carrying out public life tasks which socially correspond to men. Their function as both mother and father is one of the aspects which most frequently comes up in questionnaire surveys done on fishermen's wives. To cut a long story short, their situation is usually extremely delicate, with the aggravating factor of a very high mortality rate amongst fishermen, which leaves widows in dismal economic situations. These conditions are rejected by women and this has a significant influence on their decision to emigrate.

Globality and the exhaustion of marine resources

Today we could say that the question of the duality between globality and locality, which characterized many seafaring communities, is curiously decaying. This does not mean it doesn't exist any longer, but it manifests itself in a different manner. The improvement in transport links and communications, as well as emigration, opened and widened the framework of social relationships. In this way, that which is local is not understood in the sense of an enclosed community, but as something indigenous which must be preserved as part of a culture which starts at the level of personal experiences and reaches the identification of Galician identity as their own. On the other hand, that which is global is now part of everybody's common knowledge, at least when talking about the developed Western World. It has therefore lost its significance as human and visual contact with other cultures providing people with experiences, which helped broaden their horizons.

Nowadays, we can still consider many of these communities as enclosed, with over 80% of their resident population born in that same municipality. But, on the other hand and in spite of the crisis, there is still a representative part of the population working on ships that make their catches far off the Galician littoral. The Seychelles, Argentina, New-foundland and Namibia are some of the countries frequented by Galician fishing boats. Different current agreements usually require part of the crew to be from the country with jurisdiction over the waters, which in turn entails a greater contact with the culture of that country. We do not refer to mere sporadic relationships connected to harbours, but longer periods of human contact. In any case, one must consider that there are evident gaps, either of linguistic or cultural nature, such as happens with the agreements with Morocco.[4]

We were saying before that one of the causes of the internationaliza-tion and the globalization of fisheries was the improvement of technology that allowed the exertion of a heavier pressure on resources, while demand for fish was increasing. The exhaustion of the fisheries near the coast created the need for new, ever-more-distant fishing grounds. Our little scientific knowledge of the seas and the belief in the inexhaustible riches of fishing stimulated an over-exploitation that is now starting to be seen as noxious, but which nevertheless still continues today. Initially the lack of regulation over waters 3 or 6 miles beyond the coastline allowed total freedom over species, without respecting sizes or quantities. As the 200-mile area of exclusive economic interest was gradually being imposed, the situation changed for the better, although still plagued with defects such as the scarcity of inspections and several cheating strategies. In our interviews with seamen, they themselves recognized the abundant irregularities incurred throughout the years ranging from the existence of hidden freezing chambers to the use of nets with an illegal mesh size.[5] Not to talk about the enormous wastage derived from the non-sellable species rejected and thrown back, which can represent an important percentage of catches.

In any case, the expulsion of ships from areas of exclusive economic interest did not put a stop to either globalization or to the exhaustion of fisheries. In search of alternatives, countries with political or economic problems were targeted; such is the case with Namibia, more concerned about war than about its own waters.[6] In this respect, many current agree-ments with African countries allow access to fisheries in exchange for monetary payments; too often, the countries' poor technological levels make it difficult to carry out an effective surveillance of the catches, this resulting again in over-fishing. Other formulae used in this binomial 'internationalization-predation' are the ships sailing under a flag of con-venience, seeking shelter in those countries which have not signed international agreements and which usually act in an unethical way. There is also the question of the fishing of species going on large migrations, caught in international waters, outside areas of exclusive economic interest; there are normally international organizations which regulate these catches, but the efficacy of their inspections and their system of sanctions leave a lot to be desired.

Conclusions

The importance of fishing in Galicia from the cultural, social and economic points of view, has caused it to be given special treatment. Contradictions between Atlantic Galicia, Mediterranean Spain and Continental Europe, have been resolved against Galicia, as it is much weaker. The end of the fishing agreement with Morocco (November, 1999) does nothing but highlight this fact: important interests for the Galician economy are pitted with the apprehensions of Spanish Mediterranean agriculture, fearful of the facilities that could be given Morocco (in exchange for the fishing) to export its fruit to the Community. On the other hand, Continental Europe considers the effort of a new fishing agreement with Morocco too expensive and does not show major interest in it; The presence of an Austrian commissioner for agriculture and fishing and of a Dane at the head of Direction-General XIV (Denmark has fishing interests which are completely different from Galicia's) does not make the situation any better.

The crisis of fishing in Galicia is now very apparent, both in small villages and in large ports. The exhaustion of its own coastal fishing grounds and later the internationalization of fishing had significant consequences on small settlements. A growing rejection of the profession of fisherman adds to the need for infrastructures and facilities appropriate for the unloading of large amounts of both fresh and frozen fish, as well as for its commercialization. These lacking, there is a concentration of industry only around certain ports, whereas the smaller ones are suffering from this crisis more acutely. These survive on the scarce catches they get from coastal fishing and as providers of labour. Even this labour tends to migrate to big ports, where there is more employment opportunities. Therefore, we are witnessing the disappearance of their main economic activity, male emigration and above all, female emigration. To sum up, the dismantling of small fishing communities, above all of the most isolated ones.

On the contrary, most of the unloading and the industry is concentrated in large ports and some towns. Their main concern lies in the lack of fishing grounds and in the continuous need to find alternatives to the ones already existent. Growing economic and strategic interests around fishing as well as the over-exploitation of resources allow us to forecast a future full of uncertainties for this sector that is of such fundamental importance for Galicia. We should not ignore our own responsibility in the exhaustion of our own fisheries, a responsibility we share with other countries, which is still no mitigating factor. Small local communities were not able to maintain their own resources and the globalization of the sector is putting an end to them in favour of cities. But cities, which led the process, did not act with moderation either and today these abuses, as well as a false concern about the state of the seas,[7] are being used as justification so that those countries at the core of European economy can take advantage of peripheral ones.

Notes

1   In Galicia alone between the years 1989 and 1996 there were 241 deaths and 449 people seriously injured in accidents at sea, which represents 20 to 25% of the total number of work accidents.

2   The remuneration of fishermen is a terribly complex issue. At present, only 6% are protected by industrial agreements with a salary not exceeding 10.000 Euro per year. Although it is true that they also get a share of the catches, their small base salaries have consequences for them when it comes to getting a pension on their retirement. For the remaining 94%, including a big number of fishermen on large freezer-boats, situations vary greatly, but they are always very precarious being at the total discretion of the ship-owner, and depending on markets and catches. In any case, a retirement salary is not at all guaranteed for them.

3   Zulaika (1993) in a survey done on fishermen's wives found out that most of them, as far as possible, would not like their daughters to marry a fisherman; besides, over 50% were unhappy with their men's work.

4   Some small friction is common between Galician and Moroccan fishermen. For example, North Africans are sometimes accused of not having the right skills for the profession. More frequently, arguments centre on religious issues, either because of forbidden foods or during Ramadan, when their performance at work decreases considerably.

5   Even today, in spite of a substantial increase in surveillance, there are some irregularities such as fishing in forbidden areas or the use of two nets superimposed, the larger-meshed one hiding the smaller one that catches the smaller fish.

6   In fact, rich Namibian fisheries were nearly plundered and one of the first measures imposed after the country's independence was to closure of its waters. This entailed the expelled ships having to look for new fishing grounds on an already characteristic ship migration.

7   Very often, environmental discourses conceal economic interests. In the case of fishing, these are used as an excuse to order false biological halts or to expel a certain fleet from a fishery.

References

Calo, F. (1996), *Xentes do mar. Traballo, tradición e costumes*, Biblioteca A Nosa Terra, Vigo.

Europa en cifras (1992), Eurostat, Luxemburgo.

Fisheries. Yearly statistics (1994), Eurostat, Luxemburgo.

Giráldez, J. (1996), *Crecimiento y transformación del sector pesquero gallego* (1880-1936), Ministerio de Agricultura, Pesca y Alimentación, Madrid.

González, F. and Macau, J. (1996), *Fortaleza y debilidades de la Europa Azul*, Fundación Caixa Galicia, A Coruna.

Pinheiro, A. (1997), *As mulleres en Muros. Un estudio desde a Xeografía do Xénero*, Universidade de Santiago, Santiago.

Santos, X. (in print A), 'La crisis del fletán negro. Sus participantes vistos desde la perspectiva de la prensa canadiense', *Estudios Geográficos*.

Santos, X. (in print B), *La explotación de los recursos marinos: pesca, acuicultura y marisqueo*, Universidade de Santiago, Santiago.

Varela, M. *et al.* (1997), *Impacto en el sector pesquero gallego de las nuevas tendencias comerciales*, Fundación Caixa Galicia, A Coruna.

Xunta de Galicia (1993), *La política pesquera de Galicia en el contexto comunitario*, Xunta de Galicia, Santiago.

Zulaika, J. (1993), *As distancias do marinheiro galego en Terranova*, Lindeiros da Galeguidade II, Consello da Cultura Galega, Santiago, pp. 125-131.

18 Tourism and sustainable development – 'snow tourism' in European Mountain regions

FRANCESCO LÓPEZ-PALOMEQUE AND MARTI CORS-IGLÉSIAS

Introduction

This paper examines the traditional model of tourism in European Mountain regions and current changes that are occurring within a framework of sustainable development. Upland areas, spaces with their own distinct geography, provide the stage for a wide range of tourist activities. These activities are a clear reflection of the process by which tourist areas form and the imbalances in development that they manifest. State policies have been instrumental in this process of formation given the nature of tourism and the importance of the latter as a factor in development, above all in disadvantaged and mountain areas.

Today tourist activities tend to incorporate the principles of sustainability so as to mitigate the detrimental effects they might have on the environment, and as a response to demands from society at large and environmental groups. In turn, these demands serve to justify the intervention of the public authorities. The tourist sectors undergoing such changes include, among others, tourism in national parks and agro-tourism, where the aptness of this new model of tourism is being borne out by experience. Proposals have also been made to incorporate the principles of sustainability to what is known in continental Europe as 'snow tourism' (*tourisme de niege*) in which the viability of ski resorts is postulated within a model of sustainable development of mountain areas. This article seeks to clarify the underlying reasoning for and to determine the reach of such thinking.

Mountain areas, unique geographical spaces

Mountain areas as spaces with their own distinct geography are defined by a variety of criteria in relation to their geographical condition (altitude, slope, etc.) but, yet, their limits are ill-defined owing to the variations in these criteria caused by their line of latitude and the distortions activated by the functionality of the region, configured by its social and economic organisation. The inequalities and imbalances in these regions match the diversity and complexity of European Mountain areas.

The evolution in the social organisation and the spatial model of the settlement of mountain areas has been determined by a number of constant, yet interrelated, factors: that is, the parameters of the physical environment, and the cultural variables of an endogenous and exogenous nature. The progressive incorporation of mountain areas within wider economic zones, within the regional and international economies, has led to a transformation of the traditional mountain economy and its settlement pattern, as well as to the depopulation of these areas, converting many of them into regions of socioeconomic depression with all the signs of marginal behaviour. The timing of this process has been distinct in each range of the European Mountains, due to regional differences and the stage reached in the evolution of their economy (Rodríguez, 1983; Majoral, 1997).

In the last few decades of the 20th century this age-long process has taken on a greater complexity and its evolution has adopted a multifaceted character: some mountain regions are in the full throes of development, others are going through a process of revitalization, while others continue to show all the signs of marginality (Leimgruber, 1992). Consequently, due to these shifting trends, today it is possible to identify the following behaviour in the European Mountain regions:

1) A brake on the destructuralization of mountain areas as those factors that made urban areas so popular lose their attraction;
2) A consolidation and spread of the appreciation of the worth of mountain areas and their new economic functions, which includes an increased awareness of the environment and of the need to protect nature and
3) The strengthening of mountain areas as areas of attraction, in terms of number of visits and as an item of consumption both as a productive good and the setting for business.

These processes have generated the urbanization and/or the definitive consolidation of the settlement of mountain areas via new functions and productive activities. The mountain areas, on the threshold of the 21st century, have become, above all, areas for residence and for the exploitation of resources, areas of nature and culture that need protecting and conserving, with an educational function and, finally, areas of leisure and tourism.

Within this new model of development for the mountain areas, tourism has become one of the main development factors, though not one without its contradictions. The mountain areas, heterogeneous and diverse, boast a great variety and quality of resources, some of which are unique in relation to the rest of the region. The diversity of resources are being exploited to varying degrees, but overall their potential is high and the possibilities of converting the resources into productive output are growing thanks to the ever increasing social appreciation for them and the rising demand from tourism. The relatively disperse nature of these resources gives rise to a relatively widespread map of mountain tourism, with some exceptions that include the enclaves of 'snow tourism'.

The great diversity of resources also gives rise to an equally great variety of tourist activities and, consequently, the existence of specific tourist areas. Broadly speaking, there are five main products for the tourist of mountain areas: 'snow tourism' (down-hill and cross-country skiing, plus all modern variants); green tourism and eco-tourism (which includes nature reserves and national parks); tourism based on cultural heritage and the visiting of monuments; sports and adventure tourism and, finally, tourism that simply exploits the beauty of the landscape and the peace and tranquillity of the countryside. This variety of resources does not cancel out, however, the importance of 'snow tourism' in the main mountain ranges, an importance, which is undeniable given its economic, social and regional impact. We should perhaps not forget that the tourist activities linked to skiing are exclusive to the mountain areas, unlike other activities, such as green tourism, cultural tourism, and adventure sports, which enjoy the privileged backdrop of the mountains though this is not the only setting available for them.

Tourism and sustainable tourism

The opening up of the mountain areas to tourism is centuries old, although until recently difficulties of access impeded its development and limited the number of visitors coming to these areas. The origins of mountain tourism are diverse, but if we look solely at the most recent phase of its development we see that it has been consolidated within a society which considers leisure and travel to be a basic necessity. In recent years we have witnessed a growing appreciation of the value of nature; a search for a recreational alternative to that of the urban environment and the urbanized coastline and, moreover, a growing environmental awareness throughout society, something which barely existed two decades ago. The result of this new way of thinking and behaving is that modern, heavily urbanized, society seeks greater contact with nature - mountain areas are seen and idealized as the maximum exponent of this. The mythical notion of mountain areas, shaped throughout history, has not disappeared in this age of reason. On the contrary, as Martínez de Pisón (1981, p. 25), points out the age of myths continues to the modern day, more entrenched than before if such a thing were possible, based on imaginary constructs via the commercialization of mountain areas, the publicity of mass tourism, and the manipulation of men and regions.

These origins explain, to a great extent, the specific motivational factors of tourism in mountain areas, as is reflected in the many studies carried out on the reasons for tourist visits. The main reasons, according to these surveys, are: the desire to be in contact with nature; the great variety of sports; the distance from urban centres; the variety of landscapes; a well-preserved ecological environment; the sense of adventure; the attraction of seasonal changes on the landscape; the aptness of the natural resources to ensure the tourist gets full enjoyment for his time spent there; the existence of an adequate functional structure; the existence of a normally balanced and stable ecosystem and, among others, certain rural aspects such as the local customs of the villages. But, in addition, as has been shown in a study conducted in the French Alps (Bourguet *et al.*, 1992), people do not only visit mountain areas for leisure purposes as these areas fulfil essential functions for individual stability and social integration.

Mountain areas, full of myths, symbols and meaning, have not failed to attract the attention of society and these interrelationships are of such an intense nature that the authorities have opted for global policies in seeking a consensus as to how these areas should be managed (Gerbaux, 1994). The human pressure to which mountain areas are currently being exposed because of the growth in tourism and the fragile nature of these environments has led to the introduction of new interpretative concepts, such as sustainability, which serves as a theoretical and operational tool of debate for those institutions and bodies concerned with environmental conservation and for the productive system in general. According to the World Tourism Organisation, sustainable tourism, which includes tourism and the associated present and future infrastructure, can be defined as that which acts in accordance with the natural capacities for the regeneration and future productivity of the natural resources; which recognizes the contribution from individuals and the communities, the customs and life styles in relation to the tourist experience; which accepts that the people should receive a suitable share of the economic benefits; and which is guided by the wishes of the people of the place and the communities in the host areas (OMT, 1997).

'Snow Tourism' – basic parameters

The tourist development of European Mountain areas has been based on the exploitation of natural and cultural resources. The traditional model for mountain tourism includes types of tourism that are both spatially diffuse and concentrated - among the latter the obvious example are the ski resorts, enclaves based on the exploitation of snow as a resource and on a generally low level of concern for environmental matters. 'Snow tourism', as mentioned earlier, is the only one exclusive to mountain areas.

In spite of great technological progress in recent decades, ski resorts continue to depend on the quantity and quality of snow, though they also require suitable topographic conditions for sporting activities and a land-scape which is attractive to visitors (López-Palomeque, 1996). The distribution of snow in Europe is determined by the location of the main mountain systems and high latitudes, though the tourist exploitation (ski resorts) occurs in those areas which lie near the centres of tourist emission and in countries with advanced social, economic and technical systems. On the map of European 'snow tourism' the areas that stand out are the Alps, the Scandinavian countries, the Pyrenees and the Carpathians. The size of the ski resorts varies from one mountain range to another. The Alps reign supreme and constitute a vast area of international tourism (Vera *et al.*, 1997). In Switzerland the Alps mean that mountain tourism and skiing are vital for the economy and the way of life (approximately 80% of the people take part in winter sports). France, Italy, Austria, Germany and Slovenia also boast a large number of ski resorts and, in 1991, together with Switzerland, they reached an agreement to protect the Alps. The physical parameters of the Pyrenees are inferior to those of the Alps (area, altitude, quantity of snow, etc.), but sufficient to allow for the development of an area of ski resorts of a regional nature (López-Palomeque, 1996).

Winter tourism accounts for between 3 and 4% of international tourism, which represents 'ticket sales' of between 15 and 20 million. Yet it is estimated that inland tourism in the winter resorts is four times higher, indicating that winter tourism is deeply rooted in the inland market. Of the various types of winter sport that exist today, down-hill skiing remains the most popular sport with more than 70 million skiers world-wide, half of which are in Europe, making it the world leader in terms of skiing resorts. The current availability for accommodation in the Alps stands at around 5 million beds (Fernández, 1996, p. 52) and it is calculated that 100 million people visit these mountains each year (about 40 million holiday tourists; 60 million weekend trippers). The French Alps boasts 30 ski resorts with more than 20 ski lifts and, in total, offers 4,143 mechanical ski lifts, making it the country with the highest number of such lifts in the world (Table 18.1). The other alpine countries are not far behind in terms of areas suitable for skiing and countless snowfields.

Table 18.1   Resorts, mechanical ski lifts and skiers in Europe, 1996

Country	Resorts	Lifts	Skiers (million)
France	431	4.143	56,0
Austria	516	3.473	43,0
Italy	260	2.854	37,0
Switzerland	480	1.762	31,0
Germany	225	1.670	20,0
Sweden	340	950	12,0
Norway	210	405	8,0
Czechoslovakia (former)	300	1.500	4,8
Spain	27	294	4,3
Yugoslavia (former)	21	200	3,0
Finland	117	505	2,0
Poland	15	110	2,0
Andorra	11	64	1,5
Bulgaria	7	50	0,6

Source: Hudson (2000)

Given its dependence on snow, this type of tourism is characterized by its marked seasonal nature and high spatial (upland mountain areas) as well as temporal (winter months) concentration. To counter the highly seasonal nature of ski resorts, in the summer they are converted into perfect settings so that the lovers of high mountain areas might take part in their favourite sports. Thus, 40% of the nearly 50 million tourists and 150 million day or weekend trippers visit the Alps in the summer.

In short, tourism remains active in the ski resorts (mountain tourist centres) and during the summer the ski lifts remain operational so that visitors can take part in mountain biking, trekking and other recreational and sporting activities. The result of breaking the dependence on the season of snowfall has resulted in the offer of summer holidays to complement those in the winter.

Winter tourism began in the Alps at the end of the 19th century and skiing was to become popular in the 1920s (Keller, 1999). In the 1950s skiing became widespread throughout the Alps and the first ski resorts opened with a considerable offer of hotel accommodation. Following the Winter Olympics of 1962 in Innsbruck, skiing established itself as a tourist industry. In the 1960s and 1970s skiing reached a high level of development and became widespread in most of the European mountain ranges, which resulted in a building boom of large hotel complexes. The greatest offer of hotels and ski resorts is to be found in the Alps so that Switzerland went from 250 such complexes in 1954 to more than 1,900 in 1990 (Antón, 2000).

The introduction of tourist activities such as skiing has acted as a factor in the development and modernization of many mountain areas. This is due to the volume of capital, which these resorts move, and, in particular, because it is a form of mass tourism which produces high levels of income. On the one hand, it constitutes the basis of the economy and, on the other, it is seen as a strategy for economic shake-up and revitalization. The development of skiing has also meant the generation of processes of tourist urbanization of a residential nature, particularly, in Switzerland, France and Italy. In Switzerland, Germany and Austria the building up of winter tourist resorts has been carried out primarily as an expansion of pre-existing rural settlements, however, in France, this development has occurred largely with the construction of resorts from nothing at high altitudes. In all cases, as far as the transformation of Alpine areas are concerned, this type of tourism is characterized by the building of major residential centres, the increasing importance of large multinational companies and a limited environmental awareness. If we examine the historical evolution of the spatial creation of ski resorts, four generations can be identified in accordance with the formal and functional characteristics of each process. They have each had a series of social, economic, regional and environmental ramifications (Fernández-Garrotte *et al.*, 1989; López-Palomeque, 1999).

An activity such as skiing requires an intensive land use due to the urbanization of the area and the seasonal concentration of demand, and its displacement across the *pistes*. If we take into consideration the extreme fragility of most mountain areas, observable at a range of scales, skiing undoubtedly has a number of detrimental effects on the environment. Today, skiing attracts, more than ever, the interest of private and public institutions, who seek to evaluate the current state of this activity and to plan the future development of these tourist centres. This concern is illustrated by the celebration in 1998 of the First World Congress on the Tourism of Snow and Winter Sports, held in Andorra. Following an historical and current analysis of skiing as a tourist activity throughout the world, the main aims of the meeting were: a) to analyze the future prospects of the tourist market of snow and winter sports (trends in offer and demand) and b) to establish planning strategies and policies for the ski sector. One of the basic issues for discussion was that of the policy to adopt, summed up by the following question: What measures should be adopted to ensure sustainable development in the tourism of snow and winter sports? (Keller, 1999).

The Congress dealt with the sustainable development of 'snow tourism', which has to be undertaken so as to respect the environment and to make it compatible with other more traditional mountain activities (agriculture, livestock farming and traditional crafts). In fact, the model of 'snow tourism' seeks to fulfil a variety of objectives, among which are those of an economic or financial nature given the rate of investment and the origin of capital. This results in strategies, carried out by the agents concerned, that are contradictory to (or not totally in agreement with) the general interests of the local communities and with the model of sustainable development for mountain areas. For this, and other reasons, the role of the public authorities has acquired importance. One of the aims at a policy level is that regional and municipal governments act as a platform for the promotion of this tourist sector. In this case, the State is responsible for introducing legislation that sets out the principal guidelines for 'snow tourism' and, at the regional level, a balance is sought between the interests of the visitors and those living in the area.

Similarly, regional planning policy, at state level, has guaranteed in many European countries minimum levels of protection for the landscape and the environment, protecting areas that were susceptible to tourist exploitation; but it has not been able to put a brake on the massive growth of many winter tourist centres (Figure 18.1).

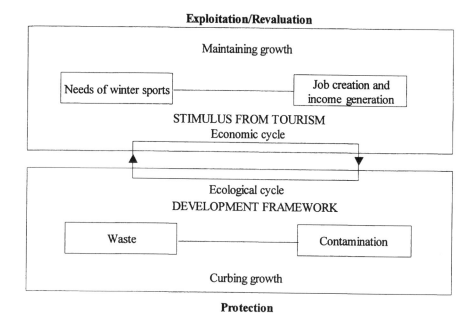

Figure 18.1   Sustainable development of 'snow tourism': The role of governments

Source: Keller (1999)

Indeed, 'snow tourism' has triggered off a series of detrimental environmental effects, in the main associated with skiing, and which have made themselves manifest at a range of levels (Debardieux, 1995, p. 85) In general, it is recognized that the human activity introduced in recent years related to skiing threatens the balance of a delicate ecosystem. Examples of this are reported in the Dobris report (EEA, 1995), some of the most significant being:

1) The substantial deforestation that has occurred in high mountain areas with a consequent increase in the risk of avalanches.
2) The visual degradation of landscapes, since the natural barriers of the forests have been replaced by plastic or wooden barriers.
3) The loss of habitats and the disturbance of species threatened with extinction because of the operating of ski lifts and the use of vehicles.

4) The contamination of water, caused largely by large quantities of sewage. For example, in the French Pyrenees the sewage from the tourist complexes goes directly into the streams and contaminates water supplies.
5) Atmospheric pollution caused by he increase in the emission of gases by private vehicles. In Switzerland 70% of tourists travel to their destination by car.

Skiing is a commercial activity. Given the importance of this type of tourist activity for the economy of the mountain regions, the State needs to provide a regional policy by introducing regulatory guidelines for the ski resorts. Within this context, the application of the principle of sustainability to the local plan is of prime concern. In policies of regional planning and environmental protection, the State needs to create the best conditions under which the winter resorts can be exploited and developed, but at all times respecting the environment. But equally it should seek to guarantee long-term demand and to widen the offer of the infrastructural base (Keller, 1999).

Conclusions

Today, most of the European Mountain areas have consolidated their position as tourist regions within the generalized spatial distribution of tourism. In this geographical setting various types of tourism are being developed, which respond to a greater diversification and specialization of the tourist offer and demand. However, within this wide range of possibilities, only 'snow tourism' is exclusive to the mountain areas and its dynamics of development and functioning confer upon them a special identity. Europe is the world leader in this sector and each year its mountains receive more than half the world's demand for skiing holidays. Within the European mountain ranges, the Alps are a leading tourist region thanks, in the main, to its exploitation of snow.

The very nature of this sector results in a complex system of economic and social interests, which today are set against the guidelines laid down to conserve and protect the environment, and even more so given the fragility of the mountain ecosystem. 'Snow tourism' finds itself debating between the demands of economic development and the need for the protection of the environment and the landscape. Elements such as the market, demand and climate, which determine the prospects for this tourist sector, are factors, which undergo considerable modifications and are usually affected by great uncertainties.

Thus, the current challenges facing this sector are to solve the typical dysfunction of a highly dynamic activity which is dependent on one resource, snow, the quantity and quality of which vary greatly from one year to the next; and to plan its future development within the framework of sustainability. According to P. Sauviain (1999), if skiing is to be expected to contribute to the sustainable development of mountain areas respecting the natural environment and the various productive outputs of the mountain areas, it will be necessary to find the optimum point between three poles; ecology, economy and society. In putting this into practice, the local communities will have to play the leading role in the drawing up of norms to establish this optimum point and in their practical implementation under the protection of the public authorities.

Yet the reality is somewhat different and in the planning of 'snow tourism' economic interests have the upper hand over environmental aspects and, thus, we are witnessing the inappropriate development of ski resorts in many countries, which are threatening the environment in general. On the other hand, there are also an increasing number of ski resorts concerned for environmental issues and, thus for example, a number of centres in the Swiss and Austrian Alps (Zermatt, Riederalp, etc.) restrict private vehicle access so as to reduce contamination.

The present development model for snow and mountain tourism needs to take into consideration, therefore, the complementary nature of commercial, seasonal and regional aspects of the skiing industry with other types of tourism, and moreover, its integration within the framework of sustainable development given recent social and economic appraisals.

References

Anton, S. (2000), 'Actividades y espacios turísticos: hacia la sociedad postindustrial', in F. López Palomeque (coord.), *Geografía de Europa*, Ariel, Barcelona, pp. 357-377.

Bourguet, M. *et al.* (1992), *Practique de la Montagne et société urbaine*, Co-édition Les Dossiers de la Revue de Géographie Alpine Hégoa, Cahiers du CRISSA, Grenoble.

Debardieux, B. (1995), *Tourisme et Montagne*, Economica, Paris.

EEA (1995), *European environment. The Dobris Assessment*. European Environment Agency, Copenhagen.

Fernández Garate, L.A, Fernández-Trapa de Isasi, J. and Fernández-Trapa de Isasi, T., (1989), 'Esquí en los Pirineos. Historia para un futuro sin fronteras I, y II', in *Estudios Turísticos* 104 (1989), pp. 101-105 and 105 (1990), pp. 79-99.

Fernández, R. (1996), *El país frágil: Las montañas deben sobrevivir*, Desnivel, Madrid.

Gerbaux, F. (1994), *La Montagne en Politique*, L'Harmattan, Paris.

Hudson, S. (2000), *Snow busines. A study of the international ski industry*, Cassell, London.

Keller, P. (1999), 'Estudio global del turismo de nieve y de deportes de invierno' in *Primer Congreso Mundial de Turismo de nieve y de deportes de invierno* (Govern d'Andorra y Organización Mundial de Turismo, Andorra la Vella 16-18 de abrtil de 1998), pp. 41-72, OMT, Madrid.

Leimgruber, W. (1992), 'Man and the Mountains. Mountains and Man: The Changing Role of Mountain Areas', in O. Galde (ed.), *Spatial Dynamics of Highland and High Latitude Environnments*, Appalachian State University, Boone, pp. 41-72.

López Palomeque, F. (1996), 'Turismo de invierno y estaciones de esquí en el Pirineo Catalán' in *Investigaciones Geográficas* 15, pp. 19-39.

López Palomeque, F. (1999), 'Turismo de montaña y nieve en España', *La actividad turística española en 1998 (Edición 1999)*, pp. 499-511, AECIT (Asociación Española de Expertos Científicos en Turismo), Madrid.

Majoral, R. (1997), 'Desarrollo en áreas de montaña', in *Geographicalia* 34, pp. 23-49.

Martínez de Pisón, E. (1981), 'Los conceptos y los paisajes de montaña', in *Supervivencia de la Montaña (Actas del Coloquio Hispano-Francés sobre las áreas de Montaña)*, pp. 21-34, Ministerio de Agricultura, Madrid.

Organización Mundial de Turismo (1997), *Guía práctica para el desarrollo y el uso de indicadores de turismo sostenible*, OMT, Madrid.

Rodríguez Martínez, F. (1983), 'Notas sobre la crisis y las posibilidades de desarrollo de la montaña mediterránea andaluza: el caso de Sierra Nevada', *Cuadernos Geográficos de la Universidad de Granada* 11, pp. 267-281.

Sauvin, P. (1998), 'El desenvolupament sostenible d'una estació de neu', in *Primer Congreso Mundial de Turismo de nieve y de deportes de invierno*, OMT, Madrid, pp. 291-297.

Vera, F. *et al* (1997), *Análisis Territorial del Turismo*, Ariel, Barcelona.

Acknowledgements

This paper has been prepared as part of the research project entitled *Delimitación y análisis de las áreas marginales en Cataluña*, funded by the Dirección General de Investigación Científica y Técnica (DGICYT) of the Ministerio de Educación y Cultura (Research Project: PB95-0905), and by an *Ajut de Suport a la Recerca dels Grups Consolidats del II Pla de Recerca de la Generalitat de Catalunya (Grup de Recerca d'Anàlisi Territorial i Desenvolupament Regional, 1997SGR-00331).*

19 Rural tourism and new patterns of development in the Portuguese rural space

JOÃO LUÍS FERNANDES AND FERNANDA DELGADO
CRAVIDÃO

Value systems and territorial development at the end of the millennium

Despite technological achievements, material gains and advances in productivity, mankind is living through times of uncertainty and insecurity in an age when concepts, attitudes and values are being redefined. This reconceptualization is revealed in the organization of territories, which thus come to be perceived and consumed in a different way from that of pre-modern or even modern societies.

In the same way, in the ambit of the increasingly complex social framework surrounding post-modern Man, policies of regional development have been restructured, concepts of development discussed and methods of intervention in territories rethought.

Until recently people believed in the success of the so-called functionalist and diffusionist patterns of development (Moreno and Moreno, 1998), here associated exclusively with the economic growth originating in urban and industrial territories; however, post-modernity has arrived to open the way for the valorization of other strategies. The concept of capitalization of low-density areas holds that, even in territories with virtually no power of polarization (according to the criteria of economic rationality), bases for an alternative, but qualitative, development may be found. Such development may also be conciliatory vis-à-vis more rationalist paradigms, and the concept itself constitutes a platform for reflection that paves the way for important debates, and hope for the future of some territories and populations. The reaction to the post-World War II models of economic growth was not, however, a novelty of the post-modernity of contemporary western societies. In the 1960s, there were already groups who fled the cities for the countryside in a movement of 'returning to nature' and the construction of supposedly self-sufficient subcultures. It was a phenomenon of reaction, which erupted with [the events of] May 1968 (Castells, 1976).

In spite of this, the current context of growing territorial interdependence we find that territorialist propositions of development have gained prominence, as if to emphasize each territory's capacity for survival on the basis of its specificities, which is the same as defending particular policies and interventions for different spatial contexts.

The practices of Fordist massification nowadays live together with the valorization of diversity, creativity and innovation. Today, short-term immediatist policies come into conflict with more far-reaching interventions. In general, development continues to be associated with urban and industrial spaces (albeit with increasingly different and more complex morphology and physiology), but now prominence is also given to alternative projects for the affirmation of territories which, in the case of Europe, have suffered a steady process of depopulation since the middle of this century. Thus rural development has gained a new dimension. This evolution is being accompanied by the emergence of new players in the development arena; from those arising from a centralized system, which usurps local initiatives, to the emergence of a dense web of interveners in the geographical space. No longer a practice of development confronted by a passive population, the inert object of external policies, but one for (or with) a population stimulated by the spirit of participatory citizenship, by a critical attitude, by the capacity for self-valorizing and self-governing their personal and collective lives. According to the principle of subsidiarity, each problem should be resolved on the scale most suited to the general welfare. Meanwhile, another concept is gaining ground that of local development, in whose ambit bottom-up decisions and interventions finds favour.

In post-modern societies, ways of becoming integrated into the group have also changed. Relationships and even the very concept of work have altered. The introduction of teleworking as a probable modeller of territorial organisation and the already visible expansion of the time given over to non-productive activities are proof of this (Umbelino, 1999). In the same way (though not on all scales, nor in all territories or social groups) the relationships between Man and resources, between society and nature, have also changed. From policies of exploitation (observed in many phases of mankind's History, and which can still be seen in many parts of today's world Geography) to new attitudes of respect for the environment. There exists an ideology of a return to nature, which first gained support among the middle and upper classes, especially in the more developed countries, and is passing through the utopia-fashion-propagation cycle, accompanying the changing values; from an industrial society to a service society, a 'cleaner, greener' society; in short, a new Man, in a new relationship with nature (Martins, 1993).

Sustainability, a concept that originated at the beginning of the seventies, has gained in importance and media visibility. Future generations are being revealed as protagonists at the defining moment and application of a model of development. At present, the scope of sustainability is widening; it is not enough to secure the quality of life of future generations – in more thoughtful and demanding social contexts, the search for excellence in daily life is also one of the contemporary population's aims.

At the same time, the notion of 'resource' has been re-conceptualized. It has a connotation contextualized in time and space, in collective needs and available technology; 'it is a resource in a given place at a certain moment'. Nowadays, in view of the new social and territorial frameworks, the concept of resource has broadened and acquired larger components of non-materiality. Has not tourist activity, to cite just one example, also been sustained by the identities of each place and the cultural specificities of each population? In times of globalization and an apparent tendency to uniformization, the non-materiality of resources emerges as a component of the process of affirmation and valorization of the 'unrepeatable' (Martins, 1993). In certain circumstances of innovation, the symbolism of singularity can thus become a factor in progress. For José Reis, one condition for development is the condensing of interrelations, a fact possible:

> ...to the extent that the spaces to be developed valorize their cultures material cultures, from the outset, because every territory has some essential know-how – but also, obviously, symbolic cultures, which represent the more solid base of self-esteem (Reis, 1998).

Is it a hazardous act of faith to believe in post-modernity where reconciliation between different value systems is possible? Is it an excessively optimistic vision? Perhaps it is the simple verification of complexity that characterizes the social framework at this ending/beginning of the millennium in societies, which have enjoyed significant increases in terms of productivity and material well-being (albeit in a dynamism increasingly marked by socio-territorial disparities and the emergence of new frontiers).

This complexity does not imply a definitive and radical re-conversion of values, in other words, of the patterns which frame the way of understanding the theory and practice of development aimed at the goals usually associated with it; wealth, well-being, comfort, quality of life. The reconceptualizations of territory and the patterns of development have not evolved in one direction only, nor have they replaced the ideologies that have generally supported economic growth in post-World War II western societies. Value systems do not evolve through rupture, but through sedimentation in successive layers, the topmost of which is more than a simple sum of, or addition to the others. It is in this logic that Leimgruber, (1994 and 1998) defines the binomial term 'secular/sacred values', the former arising from a quantitative system of competition and deterioration, the latter based on more humanistic views of everyday experiences and affirmation of territories.

This argument has an obvious connection with the landscape and the territorial organisation of societies, and brings us back to the problems of the differences – always simplistic and incomplete – between the more dynamic territories, the core definers and polarizes of the system, and the peripheral territories (or margins of this open and interdependent system). This functionalist conception is conditioned by the simplism of over-schematic visions. This geo-human reality cannot be clearly read without appealing to the problems of development patterns and value systems. In what territorial framework and in what historical context can we consider a territory to be marginal? Marginal in relation to whom, or to what? Does not the simple search for a hierarchization necessarily imply the mistake of placing each place or each population in the same logic of development?

Furthermore, the marginality of a given territory, associated with a lack of political and economic power (not necessarily in that order) is not a disaster. Within Human Geography, certain streams even hold the view that marginality:

> ...provide[s] a strategic location- a position of strength for those with new ideas about history, destiny, society and space (Smith, 1999).

The social complexity in which we live opens the doors to other possibilities for affirmation. Thus the structuring of value systems is associated with the functionality of the geographical units which make up the interdependent global whole. In every territory there is a resource, a group of potentialities, an escape route in times of crisis, a chance of incorporation into the system. In short, there are 'positive realities to discover' (Reis, 1998). We will develop this study using the example of mainland Portugal as a base.

The territory of mainland Portugal – a brief glance, the new (?) resources

Mainland Portugal has an area of approximately 89,000 km². Situated in the extreme west of the Iberian Peninsula and at the south-western tip of Europe, the mainland territory has an Atlantic coastline of around 830 km. This localizing data has clearly marked the geographical dynamism of the country, a fact that is relevant to the geo-economic activity on which we will concentrate our analysis; tourism.

Although its territory is no more than 15% of the whole of the Iberian Peninsula, the Human Geography of mainland Portugal is complex (a heterogeneity which would increase if we added the two autonomous administrative regions – the archipelagos of Madeira and the Azores – to this study). This Human Geography in a geographical space open to the flows of globalization and in contact with the global architecture of relations, is the result of the Historical past, the way in which the country has integrated in the international system, and of the policies developed and applied to the Portuguese territories, among other factors.

Portugal has a fragile urban network and a strongly polarized territory, which is reflected in an unequal, coastal population structure. There are two particularly visible spots of density; the Metropolitan Areas of Porto, and, in particular, Lisbon, where the larger slice of economic and political power is centred.

Given the maximum width of the mainland territory (there is no point of mainland Portugal more than 218 km from the coast), the country could, as a whole, be considered a coastal geographical space. Even so, until recently the map of relative distances advised some caution in this classification. The different rhythms of development also call for a more careful analysis, to the point where the segmentation (which nowadays is more socioeconomic than based on location), between a coastal Portugal and an Interior Portugal. The former coincides with the territories of densest urban population, positioned between the aforementioned metropolitan areas, and also on the Algarve coast. The latter is another side, which, although it does not show the characteristics of rurality, which dominated the country's image until the sixties, is a territory where until recently the primary sector, was the dominant sphere of activity.

Several factors have helped to accentuate the rethinking of a system of settlement badly adapted to the context of the end of the millennium, in a large part of the Portuguese territory normally considered less dynamic. They include the strengthening of Portuguese participation in the logic of external relations, the country's membership of the E.E.C in 1986 and its subsequent participation in the vanguard group of the European Economic Union. Another important factor has been the growing permeability of the country and its traditional activities to the laws of international competition and the logic of economic rationality. The urban concentration of the population, at first in the large Portuguese and European centres, now also in the more modest urban nuclei of the interior, shows a territorial dynamic that has its roots in the recent history of the Portuguese population. Indeed, thanks to the behaviour of demographic curves and the positioning of the Portuguese population in the stages of Demographic Transition, in the fifties and sixties the demographic and economic curves were already diverging. This resulted in a lack of opportunities in life for a young active population, which therefore found an alternative for survival in the European cycle of emigration, sustained by the country's position in the international division of labour; rural exodus, agricultural exodus and urbanization – territorial processes which have left their marks on the Human Geography of the Interior of Portugal and elsewhere. For example, in 1991, the population in around 52% of the municipalities of mainland Portugal was at its maximum before 1960 (Fernandes, 1996).

This is how the Portuguese population system has been adapting to the new geo-economic realities and at the end of the millennium a large percentage of municipalities exhibit low population density (in relation to the national average) and weak polarizing power, owing to the lack of an activating urban centre. They are, in fact, municipalities with a low density of economic and political relations, greatly affected:

... by the exodus to the coastal cities and by emigration, in view of the predominance of family-based agriculture – inadequate in terms of both occupation and income – throughout the North and Central Region, and given the large-scale, more or less extensive and mechanized agriculture practised in the South, which employs little in terms of labour and is neither dynamic nor competitive' (Cavaco, 1999).

Globalization presupposes that the circulation of populations and investments will increase and the mobility of material and non-material fluxes intensify. The selective, hierarchical and competitive effect of this could be a handicap for territories less able to confront more open and interdependent territorial frameworks, However, this same globalization can provide other opportunities for progress for these same geographical spaces, albeit with different, though complementary, referential and logical bases.

The depopulation of the Portuguese rural territories occurred in areas where the human presence had been a historical constant. These territories are therefore a reservoir of material and non-material heritage. In the context of a more demanding post-modern society, which nevertheless has greater economic power for acquisition and greater mobility, could the reality of this heritage constitute a means, even though merely an alternative means, of affirmation? If we take the European reference into account, Portuguese economic development was late in coming. This fact could mean a degree of comparative advantage in terms of conservation of the rural world. However, the valorization of local development factors, even if they are traditional and bound to a certain territory, demands innovation, organizational ability, actors and valorization processes. Potential alone is not a resource if it does not have recourse to an organizational web, which allows the production of goods to be consumed either by autochthonous populations, or, in particular, by exogenous markets.

In short, one of the main theoretical debates on the potential for Tourism in Rural Space is this; How, in the context of Europe, can the Portuguese rural landscape find alternatives for development based on tourist valorization of its traditional activities, monasteries, its manor houses and its popular culture? Will these material and non-material resources be determining factors in the incorporation of these territories into the international system? Could [areas of] low density be the foundation of a diffuse tourism, which will improve the welfare of local populations? Alternatively, will this dispersed tourism be a *utopia* (Balabanian, 1999) in territories which have not found their *raison d'être* in the new geo-economic reality of Portugal, and now, owing to globalization, of Europe and the world? Will it be a solution that will serve only to maintain buildings but not take care to support a local population structure? A long debate and one to which we shall try to add some points in the lines that follow.

Development, tourist activity and rural tourism in Portugal – recent evolution

The morphology of the Portuguese territory is deeply marked by tourist activity, even on a local scale. Around 7.33 million visitors in 1990 increased to 8.75 million visitors in 1997. Its climatic characteristics, the length of the coastline and the sandy nature of much of the coastal strip have combined to make the country a tourist destination, with interest focusing especially on contact with the sea, notably in the Algarve, and on some urban centres with greater powers of attraction, such as Lisbon.

Compared with countries like France, Switzerland and Austria, the official institutionalization of the commercial figures for Rural Tourism in Portugal is recent (the end of the seventies). From the beginning, the philosophy of this activity was to motivate and settle those of the active resident population who showed more initiative in the less dynamic areas of the country:

> As has happened in other European countries, at the beginning of the seventies rural tourism began officially to be viewed as a product to be developed and commercialised for a domestic and foreign clientele with purchasing power and a desire for split holidays, some spent in spaces whose landscape is either majestic or beautiful and gentle, green and humanised, peaceful and cared for by a still-numerous resident population engaged in agro-pastoral activities (Cavaco, 1999).

Rural Tourism is comparatively insignificant (around 5% of overnight stays) relative to the total tourist activity developed in Portugal. The low level of the absolute values may not, however, correspond *a priori* to a low level of importance in the local geographies in some places in the rural space. While the absolute values of tourist activity may still be of little importance, they have nevertheless resulted in available accommodation increasing by about 176% between 1990 and 1997, and an increase of about 135% in total overnight stays.

Indeed, from the sixties Portugal saw itself as a country of potential for the so-called three S (Sun, Sea, Sand) tourism, but in some areas of its coastal geography the national territory is already showing the signs of the scarcely sustainable effects of an excessive concentration in terms of time, and especially in space, on the supply available (Fernandes and Cravidão, 1997; Cravidão and Cunha, 1996). As regards temporal concentration, Rural Tourism in mainland Portugal during the nineties was not a significant novelty, as around 70% of overnight stays were in the summer months (June, July, August and September).

However, this activity is more innovative in territorial dispersion. Indeed (Figure 19.1), Rural Tourism is diffusely represented throughout every slice of the Geography of mainland Portugal, albeit with some local variations. Values are most important in the Northwest of the country, in the Minho region and, most especially, in the area around the municipality of Ponte de Lima. This is due to several factors; the pleasant climate, the

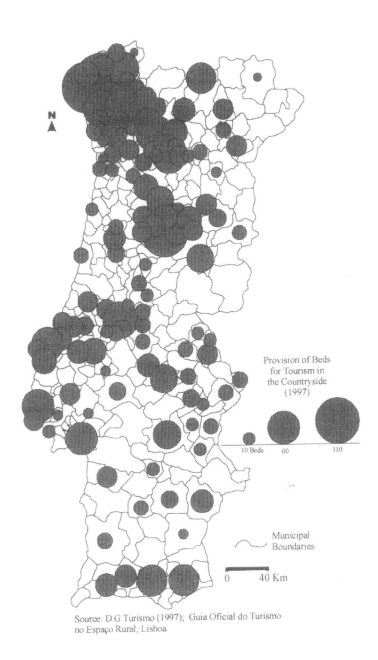

Figure 19.1   Provision of beds for rural tourism, 1997

freshness of the countryside, a more consolidated rural space, with a scattered population, and the symbolism associated with *Vinho Verde* and building heritage (especially manor houses and religious houses restored for this purpose), in conjunction with the initiative of some local associations. To lessen the effects of the fragmentation and dispersion of available accommodation, the promoters of rural tourism formed an association (TURIHAB, founded in Ponte de Lima in 1983), which provides various services, safeguarding the quality of the accommodation available, ensuring the operation of central booking offices and promoting integration in the international spaces from which a significant number of customers come (Cavaco, 1999).

We also find rural tourism units in mountainous areas such as Serra de Estrela, Gerês and the Serra Algarvia. Places with water like the Lagoa de Óbidos and the Castelo de Bode reservoir; in the centre of the country; the Douro valley and the Port producing area. In a different context, the Alentejo with its more arid landscapes indicating Mediterranean Portugal is already indicating a country of new functionalities. Use is made of old *montes alentejanos* [large estates in Alentejo] seats of former large-scale agricultural exploitation that have fallen into decay. Advantage is taken of the new accessibility, created since 1986 by the implementation of successive National Road Traffic Plans which have meanwhile spread the country from North to South and from the Interior to the Coast with a network of road infrastructures.

As this is a recent commercial activity to find expression in Portugal, there are yet no structural results on its impact. However, some studies on specific sectors of the national territory (see Martins 1993, for the case of the NW) associate the development of this activity with a 'peaceful revolution' in the rural space. If rural tourism is associated with the processes of urbanization and tertiarization, the multiplying factor of the activity seems to merit greater caution. In fact, even in the case of NW Portugal some limitations are indicated at this level. A costly supply corresponds to poor occupation rates, further characterized by seasonality, which creates problems in terms of generating employment and capital. On the other hand, at least in this case, the number of families intervening directly in this activity is small, because the entrepreneurial initiative has come from a restricted and elitist group of local players. These facts are grounds for the pessimism expressed by O. Balabanian (1999), when he says that, in spite of the combination of circumstances favouring the success of this activity in rural space:

... green tourism is not – with a few exceptions – a solution for the wasting away of fragile countryside. Green tourism is more an answer to a sociological phenomenon than it is an economic innovation. Yet the two should not be confused, nor social steps and local development assimilated.

Even so, in the case of Portugal, although always based on government subsidies, one of the areas where Rural Tourism has had a positive impact has been in restoring / conserving pockets of architectural heritage. The valorization of these could play an important part in designing local development policies, especially in increasing the self-esteem of autochthonous populations. However, what does agree with the territorialist theses of development, which indicate multi-functionality and growing flexibility for the territories, is that Rural Tourism cannot be more than a complement, a link in a chain of multi-activities. It can only have a small impact on a territory, on pain of disparaging the spirit of innovation / conservation that this activity represents. Rural Tourism is intrinsically complementary. This way of consuming territories is meaningless in the absence of farming, livestock-rearing, local crafts, and a network of good restaurants, services, in a galaxy of traditional/innovative occupations that surround this leisure activity.

Rural Tourism, therefore, can be no more than a clearing in a cluster of personalized intervention policies in the more fragile areas, whose future is open. As wide open as the roads to integration for these territories scattered within the global system. To sum up, both territories and territorial policies are merely domains for reflection, a reflection that is still to be intensified.

References

Balabanian, O. (1999), 'Le Tourisme Vert: Défi ou Utopie?', in C. Cavaco (coord.), *Desenvolvimento Rural: Desafio e Utopia*, Estudos para o Planeamento Regional e Urbano 50, Centro de Estudos Geográficos, Lisbon.

Castells, M. (1976), *Lutas Urbanas e Poder Político*, Col. Cidade em Questão 5, Afrontamento, Porto.

Cavaco, C. (1999), 'Turismo Rural e Turismo de Habitação em Portuga', in C. Cavaco (coord.), *Desenvolvimento Rural: Desafio e Utopia*; Estudos para o Planeamento Regional e Urbano 50, Centro de Estudos Geográficos, Lisbon.

Cravidão, F. and Cunha, L. (1996), 'Tourism and Sustainability: the Example of Coastal Portugal', *Partnership in Coastal Zone Management*, Samara Publishing Limited, Cardigan.

Direccção Geral do Turismo (1997), *Guia Oficial do Turismo no Espaço Rural*, Lisbon.

Fernandes, J. L. (1996), *O Homem, o Espaço e o Tempo no Maciço Calcário Estremenho- o Olhar de um Geógrafo*, Thesis for Master's degree in Geography; Faculdade de Letras, Coimbra.

Fernandes, J. L. and Cravidão, F. (1997), 'Tourism and Sustainability in Marginal Regions', in G. Jones and A. Morris, *Issues of Environmental, Economic and Social Stability in the Development of Marginal Regions: Practices and Evaluation*, Departments of Geography, Universities of Glasgow and Strathclyde, Glasgow, pp. 198-204.

Instituto Nacional de Estatística (various years), *Estatísticas do Turismo*, Lisbon.

Leimgruber, W. (1994), 'Marginality and Marginal Regions: Problems of Definition', in Chang-yi Chang, *Marginality and Development Issues in Marginal Regions*, National Taiwan University, Taipei, pp. 1-18.

Leimgruber, W. (2001), 'Globalization, Deregulation, Marginalization: Where are we at the End of the Millennium?', in H. Jussila, R. Majoral and Fernanda Delgado-Cravidão (eds), *Globalization and Marginality in Geographical Space: Political, Economic and Social Issues of Development at the Dawn of New Millennium*, Ashgate Publishing Ltd., pp. 7-22.

Martins, L. P. (1993), *Lazer, Férias e Turismo na Organização do Espaço no Noroeste Português*, Doctoral Thesis, Faculdade de Letras, Porto.

Moreno, L. and Moreno, M. R. (1998), 'Orientações Recentes de uma Geografia Social: o Desenvolvimento Local em Questão', *InforGeo- A Interdisciplinaridade na Geografia Portuguesa: Novos e Velhos Desafios* 12 - 13, Associação Portuguesa de Geógrafos, Lisbon.

Reis, J. (1998), 'Interior, Desenvolvimento e Território', *Perspectivas de Desenvolvimento do Interior*, Debates Presidência da República, INCM, Lisbon.

Smith, S. J. (1999), 'Society-Space', in P. Cloke *et al.* (eds), *Introducing Human Geographies*, Arnold, London.

Umbelino, J. (1999), *Lazer e Território*, Série Estudos, nun. 1, Centro de Estudos de Geografia e Planeamento Regional, Universidade Nova de Lisboa, Lisboa.

20 Highly specialised production as an alternative in economically depressed areas – the wines of Priorat (Tarragona, Spain)

JOAN TORT-DONADA

Introduction

Little more than fifteen years ago, the *comarca* of Priorat (in the province of Tarragona) was considered an example of an irrecoverable marginal area in Catalonia. Having suffered unremitting population loss throughout the century and with some of the worst economic indicators for the whole of Catalonia, this traditional agrarian area showed absolutely no signs of reanimation that might have allowed the future to be considered with any optimism.

Yet, a number of individual initiatives taken around 1985, aimed at reactivating, improving and promoting the area's original production (the planting of vineyards and winemaking) have brought about marked changes in Priorat. In a globalized economy that today also affects specialised production at the local level (provided high quality), the wines of Priorat are beginning to make a name for themselves on world markets, while their gradual emergence has raised hopes for development in a *comarca* which is seeking to bring its economic level on a par with that of the rest of Catalonia.

The aim of this paper is to present the main geographical features of the transformation that is taking place in Priorat. And, more specifically, it seeks to analyse one aspect in particular: the fact that this small (496 km^2) and sparsely populated (9,212 inhabitants in 1996) area is exploiting the possibilities offered by the development of its viticulture sector to shed the

261

label of marginal area and to acquire a reputation for itself in Europe and throughout the world. All this has been possible by applying a formula which, quite simply stated, involves the intelligent use of its own resources.

Priorat and the international study of the geography of wine

The cultivation of the vine is one of the oldest practices known to man, and its introduction and development are inextricably linked to the history of civilisation. Experts in botany associate the appearance of various species of the genus *Vitis* with the climatic changes that occurred at the end of the Tertiary Era, so that one species in particular, *Vitis vinifera* (which today is used for making most of the world's wines) spread throughout Europe and those parts of the world which enjoy the so-called Mediterranean climate.

Whatever the case, it is indisputable that the diffusion of this species and its evident adaptability favoured its reputation and its systematic cultivation, even since the Neolithic age. Evidence of its ancient diffusion, and of the development of a true wine culture, is to be found in the works of classical authors such as Cato, Strabo, C. Pliny and Columela, writing between the III century B.C. and the I century A.D. These authors make references to the varieties of vine, their affinities with the land and the basic principles of what would come to be known as viniculture.

With this degree of tradition, it is of no surprise that many geographers, stretching back to the last century, should have turned their attention to the significance of this cultivation, both in terms of its spatial distribution and economic importance. In recent decades, moreover, the development of the discipline has given rise to two sub-branches, *agrarian geography* and *cultural geography*, which have found in the study of vines and wines one of its richest lines of research. In some countries - France and England, in particular - this interest has seen the emergence of a specialist literature which, often, deals with questions of great geographical interest. More specifically, the relationship between the physical environment and wine production is one of the most frequently studied questions, and the debate is by no means settled (Unwin, 1997).

Information regarding the history of the area that we shall examine here will serve to place the *comarca* of Priorat in its wider context. The area owes its name to an ancient monastery, Priorat de Scala Dei, founded towards the end of the XII century and whose lands constituted the original nucleus of what is today the *comarca*. Yet the importance of this monastery does not only lie in the fact that it was an ancient feudal domain, under the protection of which a number of the villages of the *comarca* were born, but also in the fact that it played a significant economic role. The monastery was instrumental in the colonisation of these lands and in the introduction, in particular, of the vine, which with time would become its main source of livelihood. Above all, it was the XVII and XVIII centuries (in fact, a little before the onset of its decline) that saw its period of greatest splendour, coinciding with a rapid growth in the production of wines and *aguardiente* (a clear distilled brandy), which in the main were destined for export. It was also in this period that Priorat de Scala Dei acquired its reputation, even internationally, and its wines became a symbol of identity of an extensive area, stretching beyond the limits of the monastery's lands. However, the growth of this religious and economic centre was terminated here; the civil uprisings in the XIX century spelled its ruin; its lands were confiscated by the State and, eventually, auctioned off. The phylloxera epidemic, towards the end of the century, marked the end of an era and plunged wine production in the *comarca* into a deep crisis. Since then in Catalonia, and virtually right up to the present day, the name Priorat has been synonymous with a backward, marginal, economically stagnant region with a population suffering constant decline.

The *comarca* of Priorat at the beginning of the 1980s – a marginal land

Until no more than a few years ago, the *comarca* of Priorat was considered by all experts in regional planning as one of the most evident examples of marginal areas in Catalonia. A study conducted in 1974, expressively entitled *The Poor Areas of Catalonia*, included it in the group of the eight (of a total of 38) most backward *comarcas* in the region. The main factors responsible for this classification were described as: depopulation (between 1860 and 1970), the absence of industry and the low index of income per head; the final diagnosis called it 'a traditional agricultural zone with no alternative forms of production'. The conclusions reached by the economists Margalef and Tasias, in an exhaustive study undertaken in 1985, were no more optimistic: 'From the outset Priorat has to be classified as a largely isolated, agrarian *comarca* sparsely populated, with an ageing population and subject to heavy outward migration'. As for the future, the authors saw no great reasons for hope: 'An analysis of the *comarca* offers little signs of optimism for its economy, and few possibilities for the future' (Margalef and Tasies, 1985).

Whatever the case, it should perhaps be emphasised that this diagnosis was not restricted to the specific case of Priorat, but rather is applicable throughout Catalonia and Spain to many of the areas where the benefits of the economic transformation of the 1960s and 1970s had failed to arrive. This transformation had above all strengthened the industrial, trade and services sectors, at the expense of that of agriculture (and, above all, low technology, non-irrigated farming). Against this background, the economic crisis of Priorat was characterised by a number of particularly critical features: on the one hand, by the restructuring of the viticulture sector caused by the arrival, albeit at a somewhat later date than elsewhere, of phylloxera; and on the other by the general depression suffered by the whole agrarian sector, in the 1960s, at a time when industry was expanding. As a result of this process, the traditional crop of the *comarca*, the vine, which at the start of the century had acquired the character of a monoculture (21,000 hectares in 1900, occupying 75% of the cultivated land), had by 1973 lost this pre-eminence (8,200 hectares), as olive groves, nut trees (almonds and hazelnuts) and cereals gained in their share of the land.

At this point a further distinction should be made: in qualitative terms, viniculture does not have the same significance throughout the *comarca*. Based on the geological substrata, two zones can be identified: the central nucleus of Priorat, which roughly corresponds to the former lands of the monastery, and the rest of the *comarca*. The distinction is of some importance: first, because the Palaeozoic soils of this central nucleus (in which a type of slate known locally as *llicorell* is predominant) produce a wine of high alcohol content, good body and an intense red colouring; and second, because the productivity of this zone is very low, in particular because of the steepness of the terrain: from 5 to 15 hectolitres of wine per hectare, as opposed to 15 to 30 hectolitres per hectare produced on average in the rest of the *comarca*. It is of little surprise then that the crisis that has affected Priorat in the present century should have had its greatest impact on this central zone; above all if we bear in mind its mountainous orography, the difficulties -in many cases, impossibilities- for the mechanisation of agriculture, and its high degree of inaccessibility. Paradoxically, this is the zone that has given the wines of Priorat their fame and prestige throughout their history: the zone comprises some 3,600 hectares of registered vines (of which only a third were actually being used at the beginning of the

1980s) with an average overall production of some 12,000 hectolitres of wine each year. Of this, fifteen years ago, only between 20 and 30% was bottled, and the rest was sold by the litre.

In short, if required to describe the condition of Priorat at the start of the 1980s one would have to say that it constituted a prime example of a de-structured area, above all in demographic and socio-economic terms -an agrarian *comarca*, immersed in the crisis that this sector has suffered throughout the century in Western Europe, and especially in the mountain and non-irrigated inland areas. The *comarca* found itself, for the time being, incapable of escaping the vicious circle of underdevelopment, in spite of possessing resources with considerable prospects for the future, as was the case of vine growing and the making of indigenous wines.

The beginnings of a great qualitative leap – from a marginal *comarca* to a *comarca* with a world reputation for its quality wines

The situation described up to this point, though of geographical significance, is not particularly different to that of many others that have suffered the negative effects of an economic development characterised by its spatial concentration. Such development has given rise to severe imbalances between areas -at times, areas that lie very close to each other within the same country or region. Yet we have chosen to study this area in particular because of an additional circumstance: the fact that a series of individual initiatives taken in the production of wine, have given a new outlook to the economy of the *comarca*. The paradox is great: in little more than 15 years, the wines of Priorat have gone from being a product that was scarcely known and appreciated to one that has established an international reputation for itself. The facts are undeniable: in 1993, one of the labels produced in recent years, L'Ermita, became the highest priced Spanish wine; in 1997, various cases of this label were sent for sale by auction in some of the most prestigious rooms in the world: Christie's in New York and Sotheby's in London. Below, we shall examine the circumstances underlying this change of fortunes and their reasons; a change which, albeit incipient, augurs well for the future of the *comarca*.

The origin of this renewal of the viticulture of Priorat can be found in a series of personal initiatives that were taken at the beginning of the 1980s, in the central nucleus of the *comarca*: those lands, in broad terms, lying between the massif of Montsant and the Siurana river, and which coincide with the *appelation contrôlée* of Priorat -a steep terrain, difficult to farm, with very low levels of productivity: from 2,000 to 4,000 kilos of grapes per hectare (in extreme cases, only 300 kilos per hectare). Five individuals decided to put their own viticulture project into operation. Their aim: to produce high quality wines, using methods that respect to the maximum the genuine values of the product, though adopting procedures in accordance with the characteristics of the wine. In other words, it was an attempt to exploit the traditional wealth of the *comarca* but by giving it a new focus, taking as a model the best and most prestigious wines in Europe. And, above all, taking the first steps in the commercialisation process, though this meant breaking with the traditional way of thinking of the past. And, in particular, breaking the vicious circle described above: where a wine produced with such difficulties and with such great personal effort ended up being sold as wine by the litre, and used as *coupage* in wines of poorer quality.

In 1989, a few years after the project was established, the five labels produced by this group made their first appearance: Clos Dofi, Clos Mogador, Clos Martinet, Clos de l'Obac and Clos Erasmus. The following year, the group broke up, but the initiatives established were of such strength that the project continued thanks to its own momentum. Thus, new labels and even new wine producers emerged on the market: so that, since the middle of the 90s, it has become known as the 'phenomenon of Priorat' not only in Spain but also throughout Europe and the world. At the local level, although these initiatives have for the time being had a specifically localised effect, their impact on the sector has been extraordinary: for the first time in many years, the producers are able to sell their harvest at a price in line with production costs; numerous plots of land and vineyards abandoned for years are being recovered for use; and as a result of all this, the dominant trend of recent decades has been inverted; now it is difficult to find owners willing to sell their lands. Without any doubt the next ten years will show us the extent of, what we might call, this viticulture fever currently sweeping the *comarca*. For the time being, we can say that the changes that have occurred over the last 15 years, although they are perhaps not so evident throughout the whole of the *comarca*, have had a marked effect in one particular sense: that of creating, in many villages, a renewed desire to work the land and, above all, to produce wine.

Among the new wine producers of Priorat, that of Alvaro Palacios is particularly illustrative. Its owner, originally from the Rioja region, began cultivating the land in 1989. After creating the first label, Clos Dofí, he immediately set about creating two more: Les Terrasses, which saw light of day in 1990, and L'Ermita, in 1993. The latter has brought him most success on the world market. The wine has a very limited production -some 6,000 bottles a year-, made exclusively from grapes harvested by Alvaro Palacios, of the *garnacha* variety, and has received the critical acclaim of the world specialist press: the *Wine Spectator* of February 1999, claims that the 1995 wines of this label, are 'among the best wines ever produced in Spain' and was awarded the highest score that the editors of this magazine have ever given to a new Spanish wine. The establishment of Palacios in Priorat, together with others with innovative ideas for the renewal of high quality wine production, served as a catalyst for the sector. Each has adopted a method in which the key to the production success seems to lie in: the cultivation of lands in accordance with their morphology and physical characteristics; the planting of grape varieties according to the characteristics of each vineyard, disregarding pre-established percentages; the search for the greatest level of self-sufficiency in agronomic techniques, working without outside suppliers where possible and promoting their own nursery vines. The method has also sought to maintain the most important varieties of grapes in the zone (*garnacha peluda*, *garnacha lisa* and *cariñena* for the red wines; *garnacha* and *macabeo* for white wines); though this has not represented an obstacle to the introduction of imported varieties (*cabernet sauvignon*, *merlot*, *syrah*), with the aim of introducing a wider range of characteristics in the final product. Overall, it is perfectly possible to speak of a traditional cultivation, in which the human factor has played a vital role: in fact, most of the work is done by hand (in some cases with the help of mules). Herbicides and most phyto-sanitary products are avoided and organic fertilisers are preferred.

The reappraisal of the wines of Priorat – some thoughts on the possibilities of endogenous development in marginal areas

This case study of the wines of the *comarca* of Priorat, their history and the reappraisal that is currently underway, allows us to draw some brief conclusions regarding the difficulties that marginal areas face in overcoming chronic situations of economic backwardness and underdevelopment.

Clearly, the simple dualism of 'regions with possibilities of development' and 'regions without' should be avoided. The development of any region in the world depends on a great range of factors, many of which are often difficult to predict. Planning is important, but at the same time planning is not everything. The resurgence of the Priorat wines – and as a result, that of the whole region- is not attributable to planning, but rather to certain individual initiatives undertaken in a certain set of circumstances. The success of an individual initiative can act as a catalyst at the collective scale – we would do well not to lose sight of this.

In the current productive systems, typified by competition and specialisation, it is vital to get the diagnoses right and to take full advantage of the mechanisms provided by the system for the diffusion and promotion of products. It has been said that the wines of Priorat were like a rough diamond, something important was known to exist, but disguised under the appearance of ordinariness. Whatever the case, it is clear that the initial step had to be taken.

From the point of view of the intervention of the public powers, now is the ideal moment, in the case of Priorat, to behave intelligently by creating the necessary infrastructure, promoting initiatives that complement those described above, encouraging all creative undertakings, stimulating individual initiative and avoiding, at all costs, an excessive interventionism. The time is also ripe to 'copy', at other levels and in other parts of the region, initiatives of this kind. In short it is necessary to apply major injections of creativity and imagination to what is before us and which we do not always know how to exploit to the full: the authentic values, production, and culture of a place. Even when these places are remote, insignificant and unknown like this small region in southern Europe: the *comarca* of Priorat.

References

Margalef, J. and Tasias, J. (1985), *El Priorat. Anàlisi d'una crisi productiva*, Caixa d'Estalvis de Catalunya, Barcelona.
Miralles, F.; Miró, J. and Sena, E. (1974), *La Catalunya pobra*, Nova Terra, Barcelona.
Unwin, T. (1997), 'El terrer i la geografia del vi', in *Treballs de la Societat Catalana de Geografia*, 45, pp. 257-274.
Wine Spectator, 28.2.1999.

Acknowledgements

This paper has been prepared as part of the research project entitled *Delimitación y análisis de las áreas marginales en Cataluña*, funded by Dirección General de Investigación Científica y Técnica (DGICYT) of the Ministerio de Educación y Cultura (Research Project: PB95-0905), and by an *Ajut de Suport a la Recerca dels Grups Consolidats del II Pla de Recerca de la Generalitat de Catalunya* (*Grup de Recerca d'Anàlisi Territorial i Desenvolupament Regional*, 1997SGR-00331).

21 A tale right out of Hollywood – set in the desert of Almeria, in Spain?

HUGO CAPELLA-MITERNIQUE

Scouting for locations

In the mid-fifties, Hollywood and the French film industry were both on the look out for a trouble-free location for shooting their films - no problems with unions, in the case of the former, no problems with ex-colonies, in the case of the latter. In a long-forgotten corner of Spain, abandoned since the Civil War (Caparrós, Fernández and Solé, 1997), they came across an Andalucian province (Figure 21.1), immersed in great poverty, where the people's only hope was to emigrate from the accursed desert lands.

Until, that is, this oppressive sun trap, seared by drought, became the most popular outdoor film set in Europe. For a few decades the desert of Almeria was transformed into a gold mine of the film industry, into an oasis where any dream could come true. The lack of foresight shown by the local government officials, embroiled in other struggles, meant, however, that they failed to exploit this gold mine until it was too late. The crisis suffered by the cinema industry in the 1970s hit this area hard and the cinema came to form a part of a collective dream which, however, enabled a society and a land to be stirred from their slumbers. Today, the cinema continues to be a point of reference for the local economy, and it forms a recognizable trademark for the local government, which had once failed to recognize its importance.

Figure 21.1 Location of Almeria in Spain

Cinematic landscapes – fiction and reality

Behind every image, be it photographic or cinematic, lies a complex visual language in a constant state of flux. News or information becomes reality in an image, the weight of which is used as evidence of its veracity. Its supposed spontaneity, even though it might well be the product of an elaborate photomontage, suggests the capturing of a unique, unrepeatable moment. We have a magical fascination for the image and its persuasive power has been exploited in photographs that have made history, in film, in TV commercials, as evidence of extraterrestrial life, and even in science (for example, in the filming of anthropological rites). Saint Thomas, on seeing Christ's wounds, doubted no longer. Although in his case, his distrust induced him to touch Christ's flesh.

It is, nevertheless, strange to hear the criticism levelled against the wanton use of images, their manipulation, bias and subjectivity. Yet, while not wishing to detract from their power, the image remains no more than a representation of reality (Buxó and Miguel, 1999). This image or icon is transformed into a part of reality. Yet, to what extent might we say that it is a reality in itself? The social import of certain images, like the stills from the movie classics, have come to form real myths that have touched the lives of millions, although we know them to be fictitious. The depiction participates in the process of the individual perception of reality. The shadows of the cave are the reflection of a reality which we cannot perceive directly. The shadow is the image, the language, the link which enables us to express our being.

Almeria – cinematic illusions

Almeria became one of the world's main outdoor film sets. These desert lands of incomparable light changed chameleon- like to form a myriad of exotic settings. The cinema represented the beginning of the development of this province but the price to pay would be one of anonymous fame. Its bewitching landscapes would be the reflection of other lares on the world's screens.

The ravines, dried-up riverbeds, steppe-like hills, the moon-like coast of Cabo de Gata, the little whitewashed villages and the city of Almeria itself were hidden behind masks and film sets to become the fantasy locations of the movies. Fiction helped make the dreams of development come true. The fame of Almeria occurred with so muted a fanfare that most of the Spanish cinema-going public were oblivious to it. They were enraptured by the beauty of New Mexico, Arabia and Indochina, little realizing that these landscapes actually lay within their own country.

The desert lands of Almeria sweltering under 322 days of sun each year, for more than 11 hours a day (from June to September) – 8 hours in the rest of the year, with the intensity of its light, its immense stretches of deserts and rock bottom production costs (thanks to the precarious nature of the local economy) were the perfect setting for all kinds of cinematic illusions. The incomparable beauty of these places (Moya, 1998), awaited the orders of directors and the shout of 'camera and action' to participate, albeit anonymously, in the myth of cinema. Not only did they film landscapes and villages for the outdoor shots, but they also constructed sets that ran the whole gamut of scenes. Today many of these villages in Almeria are crumbling and the memories of the 'extras' who appeared in the films ring out like an echo in these primitive desert lands. Locations in the city of Almeria (such as the port and its old city square), along the Cabo de Gata (the beaches of Monsul and Genoveses), in Carboneras, Cuevas de Almanzora, Enix, Félix, Gádor, Garrucha, Gérgal, Lucainena de las Torres, Mojácar, Palomares, Pechina, Polopos, Pulpí, Rioja, Viator and Villaricos, thanks to the magic of cinema, became reflections of Aqaba, Arabia, Algeria, Bristol, Cairo, Ancient Greece, Indochina, Israel, Jamaica, Jordan, Kurdistan, Libya, Morocco, San Salvador and Turkey. In the Cortijo Ganaro (40,000 m^2) they constructed villages to film *Per qualche dollari in piu*, directed by Sergio Leone in 1965; another village was used to film *The Good, the Bad and the Ugly*, a fort in Salinetas for *Le boulevard du rhum* (1971) by Robert Enrico, a hacienda in Níjar for *Joe l'implacabile* (1966) by Sergio Corbucci, and a faithful recreation of the village of Lanoria de Arizona in Tabernas for *Valdez is coming* (1971) by Edward Sherin, are just a few of the numerous examples of the constructions that were dotted around the province, (Caparrós, Fernández and Solé, 1985).

The Far-West in Almeria

The landscapes of Almeria became one of the main sources for cinematic locations in the 1960s and 1970s. They were reshaped so as to represent a jungle in Indochina, the surface of the moon, the Holy Land and the North of Africa, but above all Almeria specialized in the filming of outdoor locations for westerns. The filming of *The Savage Guns* by Michael Carreras in 1962 was to be the first in a long line of films in this genre to be filmed in these lands until the genre was exhausted. The success of the Spanish-Italian co-production, *Per un pugno di dollari* (1965) by Sergio Leone led to a shift in co-productions which abandoned biblical themes and films set in Ancient Greece and Rome to concentrate on what was contemptuously known as the spaghetti-western. Despite previous attempts by Spanish directors, it was Sergio Leone who initiated the increasingly evident specialization of these places for westerns and which would be taken to its limits with *C'era una volta. Il West* (1968) filmed in the Leone village.

Together with other Italian film directors including Tonino Valerii, Damiano Damiani, Sergio Sollima and Bruno Corbucci, these desert lands came to be increasingly associated with the deserts of New Mexico, Arizona and Mexico. The hundreds of westerns that were shot laid down the guidelines, by which it was not necessary to stick close to reality to make the film more believable. In this way a whole set of symbols of omnipresent buildings – the saloon, the jailhouse, the boarding house, the bank, and the undertakers – was developed to represent the towns of the wild west and to make parodies of Mexican villages from which the small colonial-style church could not be omitted. Today only three of these towns remain – Leone, Mini-Hollywood and Mini-Texas – converted into tourist attractions, where shows are put on for the swarms of tourists interested in reliving the myth of the wild west. While Mini-Hollywood and Leone specialize in tourist and recreational activities, Mini-Texas, on the other hand, continues to combine this activity with the shooting of films, commercials and video-clips. Here, the summer months are dedicated to tourism while in the winter the town is handed over to the cinema. These towns are now roughly thirty years old and hundreds of films have been shot in them. They represent unique venues in the history of cinema and emblems of the western genre.

Filming on location was usually completed in a matter of days and did not always require the participation of the stars of the film. It did, however, require the deployment of a range of facilities in a place where, at the start of the 1950s, there was virtually no essential modern world infrastructure. Not all the films were shot entirely in Almeria, as often these landscapes were combined with filming in Arizona or New Mexico. In some cases sequences of Almerian locations were reused without the film crews having set foot in Spain. The Far West (in fact lying considerably further to the East) ushered in a great change, not so much economic as social, and allowed this province to break with its lethargic past.

The deserts – movie oases

In an increasingly populated world, deserts, because of their immensity and their many hours of intense light, have become the most frequently used outdoor cinematographic sets in the world. The deserts and steppe areas are rich in landscapes that have their own intrinsic interest. Not only do they offer locations for the filming of westerns, road movies, biblical, war and science fiction movies, but they also offer ideal support for the location of other quite distinct sets. The range of possibilities offered by this environment, often considered hostile for other activities, provides countless advantages for the cinema. Their marginal nature, alienating sedentary groups unable to understand them unless as a land to be colonized, is precisely what has enabled them to preserve certain redoubts of free nomadic life, away from the restrictive eyes of the law.

The film industry, as a forerunner of the globalization process of the economy and of the subsequent decentralization of activities, showed itself to be one of the first productive sectors to seek out isolated, well-lit locations. The location of the cinema mecca of Hollywood itself came about as the result of a search by a new dissident consortium from the East coast of the USA (in the borough of Queens, in New York), for a spacious and luminous place. Indeed, many cinema studios and outdoor film sets are located in peripheral steppe-like areas such as the Kalahari desert (South Africa), Texas, Arizona and New Mexico (USA) and the Rajasthan desert in India. The cinema industry saw the deserts as oases of tranquillity in which they could recreate mirages of all kinds. The isolation, peripheral and marginal nature of these lands did not constitute a problem for an economic sector which took the former nomadic inheritance of its fore-bears into show business, where travel from one city to another was commonplace.

The film industry has been a pioneer in the globalization of space. The search for remote locations has never stood in the way of making the cinema-goers dreams come true, putting lands of this kind, including Almeria, on the map.

When cinema becomes reality

The arrival of the cinema in an area such as Almeria had the effect, albeit anonymously, of offering the possibility to these lands of participating at the local scale in a process of connection with an increasingly international economy. The transnationalism of the producers implanted the seed from which local societies, for a long time marginalized from the modern world, came to play a role in it. The economic impact of these activities was highly localized, but it opened the minds of a society which at a later date was to be opened up to the world at large. Scornful expressions uttered by certain French producers, who branded Almeria as the 'arsehole of the world', incurred the wrath of local leaders who were living peacefully in their lethargy, but in turn, lit up the eyes of certain social classes, which began to see the future in a different light. The involvement of the people of Almeria as extras in the films gave them a sense of importance that allowed them to escape from their oblivion. Their faces were seen around the world and with the income from the cinema, they were able to build the much-needed well, or long-spoken-of road which would take them just a little further away from oblivion.

The dreams of the cinema enabled these lands, which in the eyes of many were of no value whatsoever, to break free. The winds of change brought with them other economic changes in the form of winter farming and tourism. The cinema backed an Almerian Hollywood and utopia became a reality.

When fiction becomes the symbol of a land

Thirty years later and when the cinematographic sector is now no more than a shadow of what it once was, the local authorities are attempting to convert it into the symbol, the trademark of the province. Since 1991, the Almerian Council, under the slogan 'Almeria: Tierra de Cine' (Almeria: Land of Cinema), has sought to publicize the genuineness of this activity and in turn recognize the key role it played for a society in search of tradition.

The fiction of these desert lands that never were, of those extras who also never were, but which provided the spark that put this place on the map, has become a landmark to which homage should be paid. But, as we all know, there is no greater threat to an activity than to celebrate its glories. The sets have become theme parks for tourists and the sole hope resides now in the shadows which can still be made out in the sets such as those of Mini-Texas, where the pioneer spirit of cinema is alive and kicking, in its wildest form. It took thirty years to rekindle a flame that long since burnt out. It took thirty years to recognize the role of the cinema as an economic and social development factor of one of the least known corners of the Iberian Peninsula.

Thanks to photograph exhibitions, a range of books and the holding of an annual short film festival the cinema has acquired a symbolic image, intimately linked to the cultural movement. Nevertheless, in the shadows and far from the mythification of the genre, landscapes continue to emerge in which the cinema-goers can continue to dream. The industry of dreams still has a presence in these places, although more silent now than ever. Yet this silence is its strongest guarantee for preserving the magic of cinema in the tranquillity of the desert.

The Tabernas desert in Almeria

The movie industry spread throughout Almeria, but became concentrated in an area of 40,000 ha known as Campo de Tabernas, or more commonly as the Tabernas desert, the only area of these characteristics in all Europe. Its extremely arid climate along with the unique qualities of its landscapes and its isolation turned this desert into the main location for outdoor filming in Europe. Today, the three villages of the wild west that remain are in the municipalities of Tabernas and Gérgal.

Under a torrid landscape

The desert of Tabernas is a small-scale inferno: annual mean precipitation is 203 mm, the climate is continental and the desert sands lie between the *sierras*: Filabres to the north and Alhamilla to the south - keeping out the Mediterranean breezes, and by the watercourse of La Calera to the east and the municipality of Rioja to the west, opening up to the peaks of the Sierra Nevada (over 3,000 metres). The desert covers 40% of the province (8,744 km2) and its rock beds comprise marls, sandstone and slate, which form a high mountain relief (Tres Picos, Jerbo, Yesón Alto) cut into by wide watercourses (including Tabernas, Lanujar, Seca, Verdelecho, Moreno), of brackish waters.

Yet despite the inhospitable nature of the badlands, the vegetation of this sub-desert habitat is highly varied in terms of the number of species (only bettered in Spain by the Canary Islands). Three main biotopes can be identified. The first is that of the hills, where the environment is arid and hostile and the only plants that can prosper are those that are xerophilous. The second is found in the ravines and gullies where there exists a marked difference between the highly salinized beds, the parched sun-facing slopes and the slopes in the shade. The latter are less arid and it is here where most of the species of this biotope prosper. The third are the terraces cultivated by man. The flora of Almeria is a link between the African and the European and contains a range of endemic xerophilous and halophilous species (Ruedo, 1982).

This landscape, because of its similarities with those of Arizona and New Mexico, was to become the new American west (Figure 21.2). However, the interest shown by the film industry was not limited solely to the Tabernas desert lands. Close by were the peaks of the Sierra Nevada,

Figure 21.2   Places related to the film industry in Almeria

with their slopes of snow, the lagoons off the coast and the sugar cane fields. A wide range of locations was on offer that could be adapted to any need.

A cinematic history

The history of cinema in Almeria dates back to the 1950s, with the first Spanish and French productions, and the area became a substitute for Hollywood, when the major American producers decided to take up residence in the 1960s. The formula of co-productions helped in the cinema boom, but by the end of the decade, because of the local crisis, the lack of infrastructure and the general crisis of the 1970s, a period of decline set in. Almeria continued to be a cinema set but primarily for Spanish films, producing a greater variety of genres and films of greater quality than in the previous decade. Today the cinema is still present, but with a greater presence in the shooting of videoclips and commercials.

César Fernández Ardarín's *La llamada de Africa* (1952) was one of the first Spanish films to be shot in Almeria, but it was the arrival of the French directors, seeking locations for their colonial war films – such as André Cayatte's *Oeil pour Oeil*, filmed in 1956 – that saw the landscapes of Almeria establish their reputation beyond the frontiers of Spain. The arrival of the great American producers, seeking cheap land, bathed in light and without the union problems then prevalent in Hollywood, began after the success of films such as *Alexander the Great* directed by Robert Rosse (1955), *El Cid* directed by Anthony Mann and *King of Kings* directed by Nicholas Ray (1960) and really took off with the making of David Lean's *Lawrence of Arabia* in 1962 and Joseph L. Manckiewicz's *Cleopatra*. The shooting of these two mega-productions meant the creation of a previously non-existent network of services in a province lacking the basic infrastructure of roads and sewers. From the roof of the bus depot in Almeria, which acted as the general headquarters, a complex operation was orchestrated whereby the local people were encouraged to organize auxiliary transport services, extras, the supply of labour to work on the sets, location researchers, and suppliers of horses and camels and accommodation for the stars and technical staff. In no time at all, a real world emerged which drove out the lethargy of a traditional society at a gallop.

The genre of films being made shifted from those set in Classical Greece and Rome, biblical stories and war films to certain subgenres, in particular, the western. The shooting of Michael Carrera's *Savage Guns* in

1962, together with the passing of legislation promoting co-productions led to the boom in the film industry in Almeria. The years in which there was a proliferation of Spanish-Italian co-productions specializing in the making of spaghetti-westerns were the golden years of cinema in Almeria: *Le pistole non discutno* (1965) by Mariano Cayano, *Finger on the trigger* (1965) by Sidney Pink, *Der Letze Mohikaner* (1965) by Harold Reini, *Per un pugno di dollari* (1965), *Per qualche dollari in piu* (1965), *Il buono, il brutto, il cattivo - The good, the bad and the ugly -* (1966) and *Nobody* (1968) by Sergio Leone. This was followed by an inevitable waning and an eventual exhaustion of the genre. The last westerns to be made had a more specifically social focus and the genre was to survive longer in Almeria than in its American birthplace (Caparrós, Fernández, and Solé, 1995).

After 1969, and the failed attempt, by the Government to attract permanent film studios ('Preferential Location Zone for the Cinematographic Industry') as well as stricter controls in the form of the granting of licenses, the film industry began to decline in Almeria, a process that was to become more marked with the local pressure and price rises caused by the absence of new projects. In the 1970s the number of films being made fell, but the quality improved and the subjects tackled expanded once more. Spanish producers turned to issues of a more local interest, but this only disappointed the local people who were by now used to big productions and film sets. However, Almeria continues to this day to be a major film set, with the filming of such films as *Conan, the Barbarian* (1981), *Indiana Jones and the Last Crusade* (1989), *Bwana* (1996) and more recently *The long kill* by Bill Corcoran, as well as many Spanish films (Fortea, 1998).

Today the film industry does not have the same social impact as it did in the Almeria of the 1960s. The stars still come but not so frequently and the local society has now developed a specific infrastructure network which does not involve the collective involvement of all and sundry. Today the legacy of the cinema is seen as a factor in the economic growth of these lands giving them an identity which has now acquired a touch of distinction, albeit at the expense of its earlier freshness and spontaneity.

Cinema as the starting motor for a marginal region

Cinema was the starting motor for the development of Almeria. It was not the result of any plan designed by the central authorities, but the result of fortune and the organization of a local private initiative which saw a golden opportunity in this new industry to open up the region to the outside world. The society, operating along traditional collective lines, was able to organize the deployment of services and activities necessary to meet the most sophisticated of demands in a very rudimentary environment.

From nowhere small transport companies, casting agencies for extras, location researchers and builders dedicated to the construction of film sets, all appeared. From nowhere hundreds of horses, camels, army squads and military equipment also appeared. From this demand, and thanks to the income generated from renting land or providing services (costumes, hairdressing), the long-awaited well for the village, the essential bridge, and the road (which in addition to allowing the lorries carrying the filming equipment to enter the region established that long-needed link with the outside world) got built.

The cinema generated new needs and uses in a society anxious for change. The first hotels and the first restaurants were built, the first tourist started arriving and an airport was opened. There was still a need for better roads, but society had set out on the road of development, the results of which we can see today. The cinema with its bohemian world of movie stars, attracted the fans, created a taste for the cinema, and allowed everyone to find a way of participating, just like the traditional celebration of the Easter processions throughout Almeria. Builders, gypsies, labourers, the region's farm-workers as well as boxers and all the eccentrics for miles around discovered a place of make-believe where the imaginary came true and cruel reality was hidden behind a muslin dress. The people of Almeria could dream and the cultural elite as in the *Afal* magazine was able to widen their intellectual horizons.

While the people dream, the authorities sleep

All attempts by the Government to promote the film industry, in general, and the development of cinema in Almeria were no more than half-hearted. On 12 June 1953, the official state bulletin announced the enactment of legislation, under the mandate of Joaquín Argamasilla at the General Management of Cinema and Theatre, regulating the granting of licenses to film in Spain. This legislation established the contractual terms for national and foreign films but above all co-productions between Spain and other foreign countries. The advantages of this latter option enabled many co-productions to be carried in the following decade.

In 1969, in order to promote Almeria, now one of the leading regions of the film business, it was declared 'Preferential Location Zone for the

Cinematographic Industry'. The measure was taken so as to endow the province a permanent infrastructure basis. Applications were invited to this end on two occasions, but both times contracts were not awarded, and the industry went into decline. The local authorities did not consider the cinema to be an essential economic activity for the province at a time when it did not appear to particularly viable, this in spite of the social ramifications it had had. They preferred to turn their attention to the traditional sectors such as agriculture, industry and tourism.

Almeria – Land of Cinema

Thirty years on, these same public authorities, following a frenetic search for a unique yet genuine mark of identity, have hoisted up the banner of the cinema. It is one of life's ironies that the shadow of the cinema has returned to centre stage, but this time in all its splendour. At a conference sponsored by the Almerian Council, the new trademark for the area was created under the slogan 'Almeria: Land of Cinema'. Plans are now underway to construct an office to manage the trademark, to establish the necessary labour conditions for new productions, to create specialized training programmes in the technical jobs linked with cinema, such as camera operators, as well as to open up permanent cinema and television studios and to prepare a retrospective exhibition on the cinematographic heritage of Almeria.

The profile of cinema is increasingly apparent in the cultural events organized in the province. Take for example the photographic exhibitions of film shoots, articles and other publications and the celebration since 1996 of a yearly Festival of Short Films.

Conclusion – the need to dream

The case of the Tabernas desert illustrates the leading role which some activities, at times considered secondary, can play in the economic and, in particular, the social development of peripheral areas. Cinema constitutes an economic sector with a global vocation, *avant la lettre*. The decentralization of the production process has meant that it has been able to seek out areas which although marginal for the development of certain economic activities, are of particular interest for cinema in that they offer space, sun and light, as well as favourable social conditions and breathtaking landscapes.

Cinema, in this case, has played a major role in a region considered isolated and marginalized. It provided the spark which allowed the dreams of a people to be lived out. It is ironic that the industry of dreams should have conjured up an image that made the lands of Almeria real. We might perhaps conclude that it is the strength of our dreams that is most instrumental in the development of a region. Dreams place no restrictions on the imagination, they are devoid of prejudices and the most fanciful of ideas can come true.

References

Buxó, M.J., Miguel, J.M. (1999), *De la investigación audiovisual*, Proyecto A ediciones: Barcelona.

Caparrós, D., Fernández, I. and Solé, J. (1985), 'La producción cinematográfica en Almería a través del archivo de la delegación provincial de cultura: 1954-1964', in *Boletín del Instituto de Estudios Almerienses* 5, Diputación de Almería: Almeria, pp. 111-133.

Caparrós, D., Fernández, I. and Solé, J (1995), 'El cine en Almería (1970-1975): el fin de una época', in *Cuadernos de Arte* 26, pp. 461-473, Universidad de Granada: Granada.

Caparrós, D., Fernández, I. and Solé, J. (1997), *La producción cinematográfica en Almería 1951-1975*, Instituto Estudios Almerienses, Diputación de Almería: Almeria.

Fortea, J. (1998), 'Pistolas en el 'Far' Tabernas', *El País -Tentaciones-* (1-05-98).

Moya, M. (1998), *Memoria del desierto*, Instituto Estudios Almerienses, Diputación de Almería: Almeria.

Ruedo, F. (1982), *Ecosistema del subdesierto de Tabernas*, Consejería de Agricultura y Pesca, Junta de Andalucía: Sevilla.

Acknowledgements

This article has been prepareda as part of a research project entitled *Delimitación y análisis de las áreas marginales en Cataluña*, funded by the Dirección General de Investigación Científica y Técnica (DGICYT) of the Ministerio de Educación y Cultura (Research Project: PB95-0905), and the an *Ajut de Suport a la Recerca dels Grups Consolidats del II Pla de Recerca de la Generalitat de Catalunya: Grup de Recerca d'Anàlisi Territorial i Desenvolupament Regional*, 1997SGR-00331).

22 Globalization and irregular urban growth from a Spanish example

MARIA JOSÉ PIÑEIRA MANTIÑÁN

Introduction

Today, the world is going through unifying expansion where a flow of exchanges, technologies, information and messages go from one side of the planet to another in a very short period of time. This is what we call a process of globalization. The main feature of this so called globalization or unifying expansion is a commercialism brought through evolution of the mass media, especially TV, which generates a desire to own properties and exploit their subsequent profits. Among the many developments that have affected the consumer society, two in particular are important. On one side, the increasing evolution of the popular fashion industry has heightened the speed of consumption, not only in terms of clothes, ornaments and design, but also on a wide range of life styles and entertainment activities, such as sports or leisure.

On the other side, there has been a growing tendency to replace goods by services, not only personal, commercial, educational or health, but also entertaining and enjoyable ones. Even though commercialization is the main feature of the globalization, this process can be analyzed as well from other points of view that play a significant role in peoples everyday life:

1) *Economic*: multinational companies are made up of branch offices and industrial networks spread all over the world.
2) *Socio-spatial*: the former antagonism city/country or centre/outskirts is replaced by the suburbs and the shopping centre, where the goods and the power of media are bound together in order to create emblematic places carefully designed to make people buy.
3) *Cultural*: social groups with different life styles spread widely; their members are given the particular areas they demand, either made for the

purpose or improved by a capital willing to maximize its profit by promoting the space fragmentation. So what we notice at first sight is a new unified localism, that is, a set of locations whose members have the same lifestyles associated with them.

In spite of that, every city pursues an urban development that may be sustainable, searching for a combination of economic and social vitality with its long-term viability, making sure the essential biophysical balance is preserved. To achieve this, it is necessary for each city to promote its productivity and to increase its competitive profits. Conditions should be created that allow people to take advantage of their city's qualities or competitive profits involved with the diversity of the local factors. These factors allow companies and industries to maintain high levels of productivity. That is the reason why there is such competence in the market of image construction, since success becomes as important as the investment in new industries.

We can conclude that nowadays the idea of a world without borders, popularized in the 1980s, is about to become a reality. However, we should be aware of its limitations, analyzing the pros and cons that it may imply. In Spain and Galicia, commercial exchanges have always been carried out but it has been proved that, if they are not clearly controlled, their consequences are an economic growth that attracts working force, and the growth of the city. This growth has caused, at the same time, its slow death since it was built with no respect for the urban morphology that identified it. The city of Vigo is a clear example of this. During the 1960s and 1970s Vigo went through a period of great economic and urban growth. However, its planning was neglected, as well as its quality of life. At that moment the industrialization produced the urbanism. Nevertheless, industrialization is nowadays shaped through the production of the urban space that expands to cover more and more population.

Vigo, the largest Galician city on the Atlantic Axis

Vigo is a city with no limits, cosmopolitan, friendly with the immigrants, a modern, working and noisy city... With little time for make-up, always busy, early riser, with much solidarity and beautiful like few cities. Vigo humbly knows it is Galicia's economic driving force and the biggest urban centre of our community (...) one of the most important ports world-wide and the door to the Atlantic ...

The above is the description of the city of Vigo (Figure 22.1) presented in the most recent tourist brochure published. No doubt Vigo is the greatest urban centre in the south west of Galicia with a strategic location on the Atlantic Axis, halfway between Ferrol and Porto (Portugal), and thanks to the Atlantic highway, distances to these two cities are shorter.

Since the end of the 18th century, the urban space development and production have always been greater in Vigo and A Coruña than in any other city in Galicia. This development has always been connected to changes in the economic activities. Both cities had remarkable fishing ports with great commercial and industrial activity that attracted working force from the country and from other autonomous regions or administrative regions such as Catalonia. The origin of the historical rivalry, still alive, between the two cities may be here, even though both cities followed different paths. While in A Coruña the industrial sector has fallen in favour of the public services one, Vigo still has a well-developed industrial sector where thousands of people work. If we intend to get familiar with the city and see how it has developed, together with its economy, in order to understand its role in the globalized space in which we live, we should analyze it from its consolidation as an urban centre until the present times.

Similar to other cities that were founded in the Middle Ages, Vigo was a small fortified centre around the Monte del Castro and San Sebastián Castle and the poor areas of Berbés-Bouzas and Valle del Fragoso. In the former there was more fishing activity and in the latter more agricultural activity. Both were the basis of the economy of the area. However, in the 1850s many people from Catalonia who worked in the salted fish industry moved to Vigo and this resulted in a demographic and commercial expansion out of the fortified centre (Pescadería, Santiago de Vigo, Areal). The proletariat appeared at that time.

New buildings were built in the outskirts of the city (Ribeira, Falperra, Salgueiral) in the first half of the 19th century. Urban activity was concentrated in Princesa (Pescadería), Constitución and A Pedra. In the 50s the subsistence economy gave way to a capitalist one. The railroad arrived in CBD -Puerta del Sol, Colón, García Barbón- and the demographic growth continued. Therefore, the walls were pulled down and new neighbourhoods were created, such as Alfonso XIII, Lepanto and Constitución. Similar to other Galician cities such as A Coruña or Santiago, Vigo had a rich bourgeoisie that demanded its own space: the Ensanches or new parts of a city. As a consequence, this social class that lived in high-valued architectural buildings carried out a land occupation-substitution process.

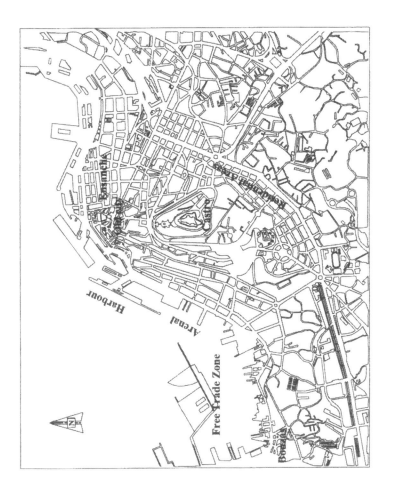

Figure 22.1   Main urban areas in Vigo

At the beginning of the 20th century, the modern industrial city took off. At that time Vigo was the main fishing-canning exporting complex. Thanks to the fish canning industry, metallurgy emerged the associated with the production of sheets, ice factories, forging, foundry, as well as shipbuilding industry of which Barreras was pioneer. In 1910 a large demographic growth took place; at that time Vigo had 35,000 inhabitants and in 1945 the population of the city had almost quadrupled (132,000), and Vigo had to annex the bordering town councils of Bouzas and Lavadores.

During the 1960s a new era began with the Stabilization Plan established by the central government, which tried to develop the Spanish economy. Its aim was an economic balance between all the regions, so that those areas in poorer conditions could be improved, for example Galicia. The creation of any type of factory and its extension and movement in the country were allowed, so the private initiative for the economic development was also encouraged. This was, however, totally inconsistent with the urban planning, especially in Vigo, a city that lacked any type of planning.

The crowding in Vigo began to increase thanks to the creation of a Free-Trade Zone, conceived with the aim of reducing customs duties in commercial exchanges, the construction of an airport, and the establishment of multinational corporations such as Pescanova and Zeltia, and the French Citroen in Vigo. The population had increased and it demanded housing. This started a new period of chaotic building that tried to satisfy the demand as soon as possible, looking for the highest benefit. As a result, builders started to build in the outskirts where urban rules were not very strict and the price of land was lower. The result was a group of neighbourhoods with big blocks of buildings of low quality and with no aesthetics at all, such as the Viviendas de Fenosa (Fenosa buildings). At the same time, a set of industrial complexes was built without taking into consideration the urban planning. The tourist industry was promoted and therefore beaches were destroyed in order to make seafronts and build hotels and high buildings, like Toralla Island, ruining the coastline.

The urban planning undertaken in the 1960s and 1970s were a complete disaster. Only the urban area, some industrial parts like Comesaña and Lavadores, and protected tourist areas like South West coastline were clearly marked. In the first one, due to the boom in the building sector, green spaces were destroyed, except for the Guía, Castro and Castrelos Parks.

The city was a group of building corridors that was far from satisfying the population. In the urban planning of the 1970s, there were sites for future buildings but they could not be started until the houses complexes were finished. Prices, as well as speculation, increased in the city centre (Pereiro-Alonso 1981). Architecturally valuable buildings were demolished in order to build others with more capacity and to obtain a higher profitability from the land. Buildings were illegally constructed. Chaos not only affected the city centre, but also the country, where single-family houses were massively built.

The picture of an economically dynamic but urbanely chaotic city

Today the Atlantic Axis, where Vigo is located, is a very important industrial space consisting of the traditional industrial centres, consolidated in the last decades (A Coruña, Ferrol, Vigo) despite the crisis they suffered in their main sectors. There are also other districts that continue with the communication of the different coast villages and the close areas. Several cities of this area are in competition to obtain sites and the industries that go out of the city, along the roads, and this results in a pattern of a continued axial industrial location (Alonso-Logroño and Lois-González 1997).

Vigo can be considered as a highly industrialized (Figure 22.2.) city despite the fact that tertiary sector has been taking jobs in the percentage of Gross Domestic Product (GDP) and in its number of employees.

Both sectors represent more than 90 percent of GDP and employees. On the one hand, from 1960 on, agriculture has accounted 18 percent less of the GDP, it represented half of the population in 1960 and declined to 10 percent of the employees (Table 22.1). On the other hand, the industrial sector was more or less balanced in the past 30 years, while the tertiary sector changed dramatically since it accounts for more than 50 percent of the working force, increasing more than 30 percent and it represents 60 percent of GDP.

Industrial activity has gone beyond the city, even the town council, and it has extended towards others close to it (Table 22.2), such as Porriño, Mos, Redondela, Nigrán, Gondomar, Salceda de Caselas and Salvaterra de Miño. All of these, except Mos, belong to the Mancomunidad del Area Intermunicipal de Vigo (Town Councils Association of Vigo) from which a joint strategic planning for their own local development is being encouraged. There, small and middle-sized businesses can be found linked to strategic sectors like granite manufacturers, automobiles, fishing, sea products or canning, shipbuilding and chemical industries.

Table 22.1   Evolution of the GDP and employment structure, 1960-1992

	GDP structure %			Employment structure %		
	Agriculture Fishing	Industry	Services	Agriculture Fishing	Industry	Services
1960	2.6	35.4	42.0	41.7	30.0	28.3
1974	10.4	40.6	49.0	22.4	34.2	43.4
1984	6.9	32.0	61.1	17.7	31.4	50.9
1990	5.1	33.1	61.8	11.4	32.0	56.6
1992	4.4	32.0	63.6	10.1	31.6	58.3

Source: López Facal, X., 1993

Table 22.2   Development of the industrial activity around Vigo

City council	Number of companies	1997 income (mil. Pta)	Variation rate
Vigo	2,417	1,224,038	30.19%
Porriño	219	108,928	20.66%
Mos	187	59,344	23.29%
Redondela	138	77,147	11.68%
Nigrán	8	15,530	34.82%
Gondomar	76	14,479	38.31%
Salceda de Caselas	63	7,728	17.39%
Salvaterra de Miño	37	4,082	36.59%

Source: Ardan, 1999

The business structure in Vigo is based mainly on the car manufacturing and assembly and industries around it that depend on how the car manufacturing develops. There is only one car manufacturing company, Citroen (Table 22.3), that accounts for more than 40 percent of the town income. This French multinational company represents 30 percent of Galician exports and has 7,500 employees.

Besides this, there are 30 other enterprises linked to it that employ 2,000 people. Undoubtedly, this is the principal driving force of the industrial sector in Vigo and one of the most remarkable in Galicia. There is a clear tendency towards a concentration of larger business complexes which

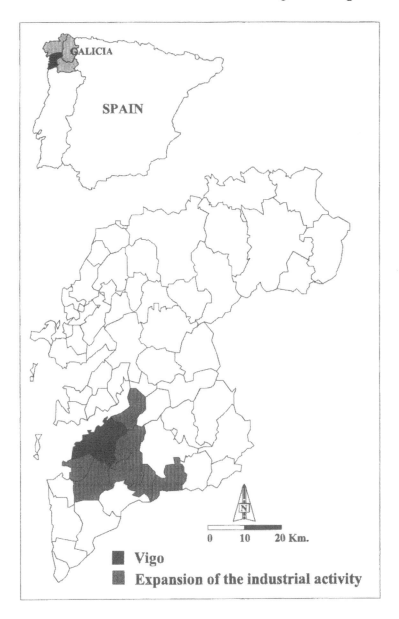

Figure 22.2　Vigo and the expansion of the industrial activity

means they depend more on foreign money. This is due to the existence of economies at different scales in some lands of car development and production, and the expansion process that has been affecting the market over the past 15 years.

Table 22.3   Economic structure of Vigo

Sector	Companies	%	1997 Income	%
Car manufacturing	1	0.03	536,798	42.05
Building ships	62	1.66	51,777	4.06
Fishing	75	2.01	37,556	2.94
Motor manufacturing	9	0.24	18,160	1.42
Hospitals	3	0.08	8,265	0.65

Source: Ardan, 1999

There are six multinationals in Spain: Volkswagen, General Motors, Renault, Ford, Peugeot-Talbot, and Citroen which produce almost all the cars and light commercial vehicles. Their strategy has been based on the search for a specialization of Spanish plants in the low-medium sector, while in their home countries, high sectors are maintained. It is strange that despite the fact that this industrial activity employs a high percentage of the working force, it is not taken into account in the urban planning. Heavy vehicles have to transport cars from Vigo to many other cities and they must go through the city centre, in the middle of the chaos due to a poorly organized network of roads. Actually, Vigo's network of roads consists of a principal artery parallel to the coast and three transversal roads, as well as some roads collecting traffic from several quarters. In relation to sea products, there is a great demand of half elaborated products that have high quality and high nutritional level, sometimes higher than those in the European Union, and it is only behind Japan.

The canning sector agrees with the investment in team goods, encouraged by current aid. Vigo is the most important port in fresh and frozen products, which has a positive effect on Galician economy.

Finally, in the ship sector, small and medium shipping fleet, represent a potential demand in the coming years which means an important increase in the rate of employment in our shipyards. The economy of Vigo is, beyond any doubt, booming with future prospects. Just walking around Bouzas or Balaidos, in the Free-Trade Zone, or going to the technological park, the goods exchange centre or the port activity park, one is impressed by the fact that there is such a large extension of land dedicated to industries and stock areas.

In contrast to this booming economy, the urban landscape has been deteriorated. Bouzas is in the city centre and Balaidos is just 3 km (1.8 miles) from the port, so arriving at Vigo by sea, the first impression one has is a group of factory premises along the port. Nevertheless, over the past few years, the population has become aware of a new image of the city. A new project called 'Open Vigo to the sea' has seen the light of day. This project tries to create·a new activity focus in the city centre seafront to revitalize that area.

The city centre is not very well-structured either. Even though, according to the Spanish Public Work Ministry, Príncipe Street is among the seven best streets throughout Europe, at night it becomes a demographic dessert, and even a marginal focus. Apartment buildings are being constantly replaced by offices. Apart from that, Vigo is a working city and people are not used to going out at night to have a drink, go for a walk or have dinner as they do in other Galician cities, after all as the saying goes: 'While Santiago prays and A Coruña has fun, Vigo works'. This also happens in the old part of the city. Some years ago, it was the business centre and nowadays there are 256 inactive premises, which is 33 percent of the total. The population in this area is primarily older, often living in poor hygienic conditions. In fact, we can say that ruin and abandon are present all over the city. Water supply is done through on antiquated network, similar to plumbing that began to work around 1950 (Lois and Cea, 1995). Very old buildings with which people from Vigo were identified have disappeared and 19th century buildings are in such a state of neglect that they represent a serious danger for the population in the outskirts and in the city centre as well. This should be more protected and looked after because it is the area that is mainly enjoyed by the neighbours and visited by the tourists.

This situation could be observed in many European cities during the last century or even today in an under-developed country. However, Vigo is not expected to suffer from it, since it has proved to be one of the most developed Galician cities and a good business network with better medium-term future prospects. The population being more sensitive to their city in order to enjoy and live in it, as well as better urban planning, may be the solution.

References

Alonso-Logroño, P. and Lois-González, R. (1997), 'Proceso de industrialización y organización del espacio en un territorio periférico', *Boletín de la Asociación de Geógrafos Españoles*, 24, pp.

Fundación Caixa Galicia (1998), *A economía galega. Informe 1996-1997*, IDEGA. Universidad de Santiago, Santiago de Compostela.

Lois-Cea *et al.* (1995), *Plan estratégico de recuperación integral y revitalización del centro urbano de Vigo*, Curso de Gerencia Pública en Administración Local. Concello de Vigo, Vigo.

López-Facal, X. (1993), *Economía e industria en Galiza no século XX*. Inédito.

Mancomunidad del Área Intermunicipal de Vigo (1999), *5000 empresas de la Mancomunidad de Vigo*, ARDAN, Vigo.

Mella-Márquez, X. M.(1994), *Plan estratégico de Vigo y su área de influencia*, Consorcio de la Zona Franca de Vigo, Vigo.

Pereiro-Alonso, J.L. (1981), *Desarrollo y deterioro urbano de la ciudad de Vigo*, Colegio Oficial de Arquitectos de Galicia, Santiago de Compostela.

Rennie-Short, J. and Yeong-Hyun, K. (1999), *Globalization and the city*, Longman, New York.

23 Sustainability, efficiency and economic success of RICA farm typologies in Italy

MARIA ANDREOLI AND VITTORIO TELLARINI

Introduction

During the last two decades, European Union has been increasingly concerned with such issues as agricultural overproduction and environmental conservation. Consequently, sustainability and efficiency of agricultural enterprises have become an important issue, both at research and policy levels. In this framework, the level of production and profitability could no longer be the only criterion for evaluating farm performance. This article analyses the performance of the farms included in the Italian Farm Accountancy Data Network (RICA) from the following viewpoints: a) economic success at farm level, b) efficiency in the use of potentially polluting inputs, and c) sustainability from an environmental viewpoint. On the base of the level of performance attained as regards economic success, efficiency and sustainability, farms have been classified into eight typologies, whose relations with structural features and styles of farming have been checked out. Although the RICA sample can not be taken as 'directly' representative of Italian agriculture, the relationships between farm typologies and their performance may, nevertheless, provide decision-makers with useful information on how to support or steer the management of farms depending on the typology they belong to.

The methodology applied

The object of this analysis is the RICA farm sample used for 1994 accountancy. This sample includes 17.292 farms, mainly concentrated in the South (43%) and North (38%) of Italy, while farms located in the Centre account for 19% of the sample. Figure 23.1 shows the Regions included in Northern, Central and Southern areas.

Figure 23.1   Northern, Central and Southern Regions of Italy

Tables 23.1a and 23.1b show the main features of RICA sample, accordingly to geographic and altitudinal location of farms. In particular Table 23.1a shows the following features of RICA sample: 1) number of farms, 2) total Utilized Agricultural Land (UAL), 3) total Gross Production (GP), i.e., total production minus the input re-employed inside the farm itself, 4) total Farm Net Income (FNI) and 5) the total amount of 'Flows'. In this article, the term 'Flows' is used for describing the amount, expressed in monetary terms, of all the inputs that cause pressure on natural resources and consequently lower their quality; both variable and fixed inputs have been included in the Flows. Variable inputs include: chemical inputs, fuel and lubricates, electricity consumption, water consumption, machinery contractors, machinery and building maintenance expenditure, etc. Fixed inputs are taken into account by calculating monetary values relating to annual costs of depreciation of farm machinery, implement and building. The amount of Flows is only a 'proxy' of the potential impact on natural resources, insofar as the dangerousness of inputs for unit of monetary value might vary considerably. However, since RICA sample provides only information expressed in monetary values for aggregate categories, it was not considered worthwhile weighting input categories according to their potential danger (Pennacchi, 1998).

Table 23.1   Main features of RICA sample in 1994: a) total values and b) average values

a) total values	Number	Utilized agricultural land (ha.)	Net farm income (million €)	Gross pro- duction (million €)	Value of 'flows' (million €)
Northern Lowland (NL)	3,646	64,276	102.45	257.92	117.98
Northern Hills (NH)	1,593	22,894	33.82	81.95	36.12
North. Mountains (NM)	1,302	38,982	21.87	57.76	25.71
Centre Lowland (CL)	983	17,055	16.85	42.33	17.69
Centre Hills (CH)	1,986	41,574	24.45	63.77	26.24
Centre Mountains (CM)	304	8,233	4.35	9.22	3.61
Southern Lowland (SL)	3,228	47,212	48.47	105.15	36.69
Southern Hills (SH)	2,641	70,529	37.76	74.64	26.63
South. Mountains (SM)	1,609	49,952	24.91	44.63	15.05
Total	17,292	360,707	314.93	737.36	305.73

b) average values	Utilized agriculture area per farm (ha.)	Net farm income (000 €)	Gross pro- duction per ha (000 €)	Value of 'flows' per ha (000 €)	Total flows/ Gross pro- duction
Northern Lowland (NL)	17.63	28.10	4.01	1.84	0.46
Northern Hills (NH)	14.37	21.23	3.58	1.58	0.44
North. Mountains (NM)	29.94	16.80	1.48	0.66	0.45
Centre Lowland (CL)	17.35	17.14	2.48	1.04	0.42
Centre Hills (CH)	20.93	12.31	1.53	0.63	0.41
Centre Mountains (CM)	27.08	14.31	1.12	0.44	0.39
Southern Lowland (SL)	14.63	15.02	2.23	0.78	0.35
Southern Hills (SH)	26.71	14.30	1.06	0.38	0.36
South. Mountains (SM)	31.05	15.48	0.89	0.30	0.34
Total	20.86	18.21	2.04	0.85	0.41

* The value of Italian Lire at 1994 has been converted in Euro using the official rate of change (1,936.27 ITL for one Euro)

Source: Our elaboration from RICA data bank – 1994

Finding indices assessing economic success, efficiency and sustainability

The concept of 'Flows' (FL) has been used for splitting the index traditionally used for measuring productivity, namely the ratio between Gross Production (GP) and Utilized Agricultural Land (UAL) in two separate indices, the first assessing sustainability and the second assessing efficiency in the use of potentially polluting inputs. In other words, the ratio 'GP/UAL' has been split as follows:

$$GP/UAL = (FL/UAL) / (FL/GP)$$

The ratio between flows and the area on which they have been employed (FL/UAL) is considered as a proxy of the sustainability of agricultural activities. It indicates how many Euro have been spent in potentially polluting inputs for every hectare of utilized agricultural land. As long as this ratio is low, it can be supposed that agriculture is relatively sustainable. The ratio between flows and gross production (FL/GP) has been considered as a proxy of the efficiency in the use of potentially polluting inputs. It indicates how many Euro of flows are necessary for obtaining one Euro of agricultural gross production. A low value of this ratio indicates a relatively efficient production (Pennacchi, 1998).

As regards economic success, gross production has not been considered as a suitable variable; consequently in this analysis farms have been classified according to the amount of net income, considering firstly the total amount at farm level and secondly the ratio between farm net income and hours of family labour. This under the hypothesis that a farm could be considered as successful from an economic viewpoint when it provides a satisfactory level of total net income and, at the same time, it provides a satisfactory compensation for family labour. Under the above-mentioned hypotheses, farms have been classified into the following groups (see Table 23.2): 1) Non-entrepreneurial-farms, 2) Unsuccessful-farms, 3) Intermediate-farms, and 4) Successful-farms.

The first group 'Non-entrepreneurial-farms' includes farms characterized by such a low level of gross production that they cannot be considered as 'agricultural firms'. Farms belonging to this group mainly consist of small plots of land, which are cultivated without economic purposes, e.g., as a hobby. The value of annual gross production under which 'farms' have not been considered as 'agricultural enterprise' was 5,164.57 € per year.

Farms with a gross production exceeding this minimum level have been considered as 'agricultural enterprises', independently from their economic success, and they have been classified as: a) unsuccessful farms, b) intermediate farms and c) successful farms, according to the following methodology.

Firstly, for each geographic macro-region, i.e. Northern, Southern and Central Italy (see Figure 23.1), the levels of net income that could be considered as low, medium or high have been defined. This was done by comparing each farm net income with the average consumption in 1994 of a three-member family for each macro-region. Farm net income has been considered low when it is less than 70% of the three-member family consumption and high when it is more than 140% and medium when it is between these two thresholds. Secondly, the ratio between farm net income and amount of family labour employed has been analyzed to assess if the farmer and its family get a satisfactory income for their labour. This ratio has been considered as:

1) 'low' (unsatisfactory hourly net farm income) when it is necessary to employ at least 4,600 hours (two family units working full-time) for reaching a low total farm net income as defined above,
2) 'high' (satisfactory hourly net farm income) when it is higher than the ratio obtained in the case that one family unit working full-time (2,300 hours) can reach a high farm net income, and
3) 'medium' in all the intermediate situations (Andreoli, 1994).

According to the class of total farm net income and the class of hourly net income for family labour, farms have been classified into the above-mentioned groups, as shown in Table 23.2.

Table 23.2   Farm typologies from a private economic viewpoint

Values at farm level		Farm Net Income/hours of family labour		
		Low	Medium	High
Low Gross Production		Non-entrepreneurial farms		
	Low	☹ ☹	☹ ☹	☺
Farm Net Income	Medium	☹ ☹	☺	☺
	High	☺	☺ ☺	☺ ☺

Legend: ☹ ☹ unsuccessful farms, ☺ intermediate farms, ☺ ☺ successful farms

From success, sustainability and efficiency indices to farm typologies

The classification of farms into typologies has been performed on the following basis. Firstly, performing a dichotomic classification based on economic results by distinguishing successful farms (S) from the other three typologies, that have been defined as Non Successful (NS). Secondly, classifying farms according to all possible combinations between level of efficiency (high or low) and level of sustainability (high or low). Since the level of potentially polluting inputs varies considerably according to the type of farm enterprises, farms have been classified into 15 groups,[1] depending on the activities (cereals, dairy cattle, etc.) that are mainly contributing to their economic results. The weighted average value of sustainability and efficiency, characteristic for each of those 15 groups, has been used as a threshold for distinguishing between low and high levels of performance.

This causes farms to be classified into the following four groups:

G1: farms that are 'less sustainable', but 'more efficient' than the RICA average for the group where they are included,

G2: farms that are 'more sustainable' and 'more efficient' than the RICA average for the group where they are included,

G3: farms that are 'less sustainable' and 'less efficient' than the RICA average for the group where they are included, and

G4: farms that are 'more sustainable', but 'less efficient' than the RICA average for the group where they are included.

Crossing the two categories nominated above as 'S' and 'NS' with the four categories nominated above as 'G1', 'G2', 'G3' and 'G4', results in the eight typologies that are presented in Table 23.3.

Farm typology main characteristics

Farm typologies, farming styles and possible future trends have been described based on the following six variables:

1) *Economic success* measured by Farm Net Income (FNI) level that is usually considered as the variable that the entrepreneur tends to maximize;

2) *Physical dimension*, measured by the farm acreage expressed in hectares of Utilized Agricultural Land (UAL) that gives information on the actual and potential level of production that the farm could attain;

Table 23.3   Farm typologies according to economic performance and level of sustainability and efficiency

Successful Farms (S)	Non- Successful Farms (NS)
G1 Traditional farms with medium-large acreage (S-G1)	Traditional farms with medium-small acreage (NS-G1)
G2 Large and efficient farms (S-G2)	Medium acreage farms whose future possibilities are uncertain (NS-G2)
G3 Inefficient and intensive medium-large farms (S-G3)	Small part-time farms (NS-G3)
G4 Large, inefficient and extensive farms (S-G4)	Area of disinvolvement and of farm disactivation (NS-G4)

Source: Pennacchi (editor), 1998

3) *Total efficiency of the farm-system*, measured by the Farm Net Income (FNI) - Gross Production (GP) ratio (FNI/GP), which measures the share of farm gross production that the entrepreneur and its family can keep as a return on the inputs (land, labour and capital) they have invested in farm activities. Consequently, farm global efficiency increases as long as the value of this ratio increases;

4) *Land productivity*, measured by the ratio between Gross Production (GP) and Utilized Agricultural Land (UAL), i.e. GP/UAL. This variable, given stated levels of entrepreneurial skills and of land fertility, usually shows an inverse correlation with the level of environmental sustainability. Moreover, it is often inversely related to the farm acreage, although this correlation may vary depending on entrepreneurial skills and goals;

5) *Environmental sustainability*, measured by the amount of potentially polluting inputs, or Flows (FL) per hectare of Utilized Agricultural Land (UAL), i.e. FL/ UAL. It measures the pressure on natural resources and it is very often inversely related to the farm acreage;

6) *Environmental efficiency*, measured by the value of potentially polluting inputs, or Flows (FL) per monetary units of Gross Production (GP), i.e. FL/GP, which measures how many Euro of potentially polluting input are needed for obtaining one Euro of Gross Production. This variable, as mentioned above, is a proxy of the efficiency in the use of potentially polluting inputs. As in the case of the efficiency of the total system, the value of this ratio depends on entrepreneurial choices as regards to re-using farm produced inputs or buying them from outside the farm.

Table 23.4 shows the value taken by the above described six variables for the eight farm typologies presented in Table 23.3; their geographical localization is discussed later.

Table 23.4   RICA sample in 1994: farm typologies main features

	FNI (000 €)	UAL (ha)	FNI/GP (%)	GP/UAL (000 € / ha)	FL/UAL (000 €)	FL/GP (000 €)
S-G1	65,053	21.17	56.15	5,471	1,780	0.325
NS-G1	14,926	8.47	47.65	3,698	1,247	0.337
S-G2	47,887	61.38	59.07	1,321	358	0.271
NS-G2	11,875	20.00	49.83	1,191	368	0.309
S-G3	61,270	35.05	37.47	4,665	2,394	0.513
NS-G3	8,043	12.53	24,90	2,579	1,469	0.570
S-G4	38,134	84.81	44,17	1,018	442	0.434
NS-G4	5,579	23.69	25,53	922	481	0.521
Average	18,213	20.86	42.71	2,044	848	0.415

Source: Our elaboration from RICA data bank – 1994

Traditional farms with medium-large acreage (S-G1)

This typology includes farms that are successful, polluting, but efficient in the use of potentially polluting inputs. It accounts for 7.49% of the total number of farms included in the RICA sample at 1994. The entrepreneurs of these farms try to make the best from the available resources and they reach the 'economic success' due to the efficiency with which they organize the farm system and to the relatively large farm acreage. Aiming to a very high land productivity entrepreneurs are forced to choose relatively polluting techniques. These farms have been named 'traditional' due to the fact that, following the traditional model of agricultural economics; they reach a satisfactory income level by maximizing the scarcest resource, i.e., land.

It would be necessary to have more information for understanding why these farmers intensify production so much, since the large acreage of their farms would allow them to attain high net farm incomes without using much polluting inputs. The intensification causes farms to be 'under stress', meaning that they produce by exploiting natural resources and external inputs to the maximum. In this scenario, it is unlikely to foresee either an improvement on productive levels or a choice to give up high net income levels in order to come back to a more sustainable style of farming. What is going to happen to these farms mainly depends on the socioeconomic context, i.e., agricultural policy as regards subsidies and input prices, land market trends and entrepreneurial goals. An increase of farm acreage, e.g., could result in reaching high income levels with less intensive techniques. In the same way, good fertility levels and climate could lower the above mentioned 'productive stress' conditions, consequently reducing the worries about the chance of negative evolution of this farm typology.

Traditional farms with medium-small acreage (NS-G1)

As in the case of the previous typology, these farms are exploiting natural resources and external inputs to the maximum. The difference in results is due to the much smaller farm acreage and to the lower level of farm-system efficiency, that do not allow these farms to reach a net income level high enough to make them being classified as successful farms. The importance given by entrepreneurs to the agricultural activity is testified by the high level of land productivity, although this is not as high as in the previous typology. Future trends are more difficult to be foreseen than in the previous case. Indeed, since in this case net farm incomes are not such as to make the agricultural activity 'viable' from an economic viewpoint, family decisions will depend on the situation of the local labour market. In other words, if it would be possible for young farmers to find alternative and satisfactory employment on the area, it is possible that these farms will assume the same characteristics as part-time farms (NS-G3 typology).

Large and efficient farms (S-G2)

As in the case of large traditional farms, large and efficient farm behaviour can be explained quite well by the traditional agricultural economics analyses. Indeed, since their large acreage, they can accept to have relatively low levels of land productivity and nevertheless reach a high level of farm net income. In this case, the more scarce resource, whose use has to be carefully planned, is no longer land, but labour. These farms have a good stock of natural resources; they use in an efficient way potentially polluting inputs and do not exploit land potentiality to the maximum. This because, although they are characterized by a level of land productivity lower than average, they can reach a relatively high level of net farm income (47,887 €). Thus, thanks to their acreage, they can afford being sustainable, although their good performance depends also on their efficiency. In this case, too, situations differ quite a lot according to the level of natural resources they are endowed with. These differences in natural resources result in differences in the level of both sustainability and efficiency on the use of potentially polluting inputs and in the level of net farm income that is attainable.

As regards future trends, it is possible to hypothesize that their large acreage would allow these farms to reach good economic results without being forced to reduce their sustainability and efficiency levels.

Medium acreage farms whose future possibilities are uncertain (NS-G2)

This is one of the three farm typologies that are more important (in terms of farm number) inside the RICA sample. The fact that it is characterized by uncertainty in future trends seems to hint that this condition is common to a large share of Italian farms. Their entrepreneurs show the typical behaviour of farmers that, due to the medium acreage of the farm, have made the choice 'to try to carry on farming', at least up to now. These farms are sustainable and efficient but do not reach economic success. They are intermediate between the situation of traditional medium-large acreage farms (S-G1) and the farms included in the area of disinvolvement and farm 'deactivation' (NS-G4). They have in front of them two alternatives; either they are able to reach higher income levels and make a living out of farming, or they will likely evolve towards farm 'deactivation', which is typical to the last typology. These three farm typologies have all similar farm acreage, but their entrepreneurs have made different choices. In this typology, as in the case of traditional medium-large acreage farms, farmers have decided to rely markedly upon the income resulting from farming, but they have decided not to exploit so much resources, thus being sustainable. Sustainability is a common feature among the farms of the 'disinvolvement' group, from which this typology mainly differs due to the higher attention given to the issue of efficiency in the use of resources. This efficiency allows these farmers to reach not negligible levels of farm net income that is much higher than that of disinvolvement group.

As regards to future trends, these farms are characterized by an unstable situation, which – as mentioned above – in the short-medium run may evolve either towards an increase of income level (probably at the expense of a lower sustainability) or towards disinvolvement and farm disactivation. The route entrepreneurs are going to take will depend from the socio-economic context (input and output prices, level of subsidies, local labour market situation), but still more from family decisions as regards their interest for farming and the level of farm net income that is considered as the minimum acceptable level. From this viewpoint, the choices of younger family members will be crucial, since they are usually more attracted than their older relatives by external and more profitable activities. Since the equilibrium of these farms is very fragile, decision-makers should be very careful in setting up interventions that risk breaking this equilibrium.

Inefficient and intensive medium-large farms (S-G3)

This typology is represented by very few cases. As for the traditional medium-large acreage farms (S-G1), the farms belonging to this typology seem to aim to income maximization. Nevertheless, they are different as far as their level of efficiency is lower than in the first typology, both as regards to the farm-system and the use of potentially polluting inputs. It could be said that since economic results are so good, these farms could afford being inefficient. Another explanation could be found in economic theory, where private entrepreneurs are considered as using marginal values and not average values in decision making. Consequently, it is possible that the maximum economic result at farm level (be it total profit or total net income) is reached when the average productivity of external inputs is quite low. In other words, an efficient use of potentially polluting inputs is an aim at 'social' level, but it could be in contrast with private economic goals. Indeed, entrepreneurs do not research solutions that give the maximum net income for units, but the maximum net income at farm level, thus taking into account both unitary net income and level of production. Nevertheless, these farms could likely improve their efficiency using some external advisory services. However, since the economic results in terms of farm net income are more than satisfactory, entrepreneurs may not have the push to look for any improvement of their actual situation. In this framework, the entrepreneurial attitude towards farming could be crucial for making it possible to improve efficiency and sustainability. Moreover, the low efficiency of the farm-system, resulting from the fact that entrepreneurs buy most of the inputs, either goods or services, might imply a scarce direct involvement of-entrepreneurs in farming activities. If in a scenario where agricultural productive levels were very important it could be justified to aim to production maximization more than to the maximization of economic results, the increasing focus on environmental problems should make farmers more sensitive as regards to sustainability issues.

Small part-time farms (NS-G3)

This typology is the most represented (in terms of number of farms) among the farms belonging to RICA sample and, very likely, among Italian farms, as well. In this typology farm net income is quite low and is not adequate for sustaining an average family if not complemented by significant out-of-farm incomes. Consequently, if an interest for farming is left, this is usually limited to technical aspects of production, rather than to economic ones. This could explain the quite high land productivity levels. As in the previous typology, the efficiency both of the global farm-system and of the use of potentially polluting inputs is quite low. Indeed, external potentially polluting inputs account for about 57% of gross production and the share of gross production retained by the entrepreneur and its family is only less than 25%. This could be due to the fact that the small acreage does not make profitable the use of own farm machinery (with the consequent use of contractors) and it makes impossible for the entrepreneur (and its family) to give prior interest to farming. In this scenario, it is natural that farm strategies aim more to optimize the global family income (and labour) situation rather than to maximize farm economic results. In other words, families try to reduce their labour involvement in farming activities to the minimum, with the aim to achieve a satisfactory per hour return on labour rather than a high total farm net income. The high use of potentially polluting inputs is also due to this wish to reduce family labour employment, substituting labour with capital, most of all with working capital.

As regards to future trends, it is possible that with the help of a good advisory service these farms will become more efficient, although – very likely – not very much more sustainable. If not, it is possible that an increasing disinvolvement in farm activities could bring them to the situation of inefficient and sustainable farms (G4). Although, from a private point of view, this can result only in a decrease of farm incomes, from a social viewpoint this evolution would be a positive one, lowering the actual high pressure on natural resources.

Large, inefficient and extensive farms (S-G4)

Farms included in this typology achieve a satisfactory level of net farm income thanks to their large acreage, although their efficiency in managing the farm-system and in the use of potentially polluting inputs is quite low. As stated for typology S-G3, these farms could afford being inefficient, due to their good level of farm income. They differ from the farms of group S-G3 due to their high level of sustainability. In this framework, since the non-efficient use of polluting inputs is not due to their excessive use (that would have caused these farms not to be sustainable), it is likely due to an irrational use of the external inputs themselves. The share accounted by large inefficient and extensive farms on the total number of farms included in the RICA sample is the lowest, 0.61%. In this typology the inefficient use of potentially polluting inputs might be due to a lacking involvement of the entrepreneurs in farm activities; indeed, they seem to be aiming more to 'land rent strategies' than to make their farms efficient from an economic viewpoint.

In this case, the high inefficiency of the farm-system in presence of a low use of potentially polluting inputs could be explained by the fact that entrepreneurs and their families are not providing farms with many inputs (first of all, their labour). Since these farms are mostly located in the South of Italy, they could be seen as a heritage of the old 'latifundia' management typical of Mezzogiorno. It is difficult to foresee the future evolution of this group of farms. Of course, as in the case of group S-G3, they could profit of a good advisory service, but this only in the case that entrepreneurs choose to be more involved in farming activities. On its turn, this depends on how much farm net income is important for the family and on which opportunities for external employment the entrepreneur and its family could have. It is important to stress that the current agricultural policy, giving subsidies on a land surface base (rather than in terms of minimum guarantee price) might encourage these forms of 'latent abandonment' of lands. It is in this direction that these farms risk to evolve if external and internal conditions able to stimulate a higher entrepreneurial involvement will lack.

Area of disinvolvement and of farm deactivation (NS-G4)

This typology is the third largest one in terms of share of farms, after the groups of small part-time farms and of farms whose future evolution is uncertain. This typology could be considered as the last evolutional step for farms whose entrepreneurs have decided not to be involved in farm activities. This disinvolvement is more a choice than a 'must' as their farm average acreage (24 hectares) could support more economic and productive ways of farm management, and the farm net income is very scarce and it is obtained by using very little input. In other words, efficiency seems not to be an issue in the family agenda, since all the income that farms are able to produce is considered as a sort of 'present', as the family is relying upon others sources than farm income. Due to this situation, it is quite unlikely that entrepreneurs will decide in the future to improve the efficiency and profitability of their farms. Indeed, they seem to have made the choice to aim more to the preservation of their capital (first of all by land) or to the minimization of their financial and personal involvement, than to the maximization of farm net income.

A general overview of the eight farm typologies

Table 23.5 shows how important are the above described typologies in terms of number of farm, utilized agricultural land, net farm income, gross production and flows. According to the data given in Table 23.5 it is apparent how the eight typologies have a different weight on the total, according to the variable that is analyzed. Thus, e.g., traditional large acreage farms (S-G1) make only 7.49% in terms of number, but they account for almost 27% of the total net farm income and for about 16% of the total amount of flows. One could expect that these differences were to be found also in the universe of Italian farms, however, since the RICA sample is not in all respects 'representative', it contains different proportions of farms in groups than the 'universe', and consequently the shares would be different. It is for this reason that the analysis presented here should be considered more 'qualitative' than 'quantitative'. A quantitative analysis in Italian terms could be performed only after weighting RICA farms. Indeed, RICA sample tends to include a higher share of farms with good economic dimension and performance. Moreover, the fact that RICA sample can not be considered as 'totally random', would ask for a higher carefulness in weighting procedures.

Table 23.5   RICA sample in 1994: farm shares in terms of number, land and economic values according to typology

	% Number	% Utilized Agr. Land	% Net Farm Income	% Gross Production	% Flows
S-G1	7.49	7.61	26.77	20.36	15.97
NS-G1	12.39	5.03	10.15	9.10	7.40
S-G2	6.92	20.35	18.19	13.15	8.61
NS-G2	21.80	20.90	14.21	12.18	9.07
S-G3	3.22	5.41	10.84	12.35	15.29
NS-G3	29.55	17.75	13.05	22.38	30.75
S-G4	0.61	2.47	1.27	1.23	1.29
NS-G4	18.03	20.48	5.52	9.24	11.62
Total	100.00	100.00	100.00	100.00	100.00

Source: Our elaboration from RICA data bank at 1994

The Table 23.6 shows how the farms belonging to each of the eight typologies are distributed among the geographical-altitudinal areas. In order to put in evidence, which are the locations more related to each typology, the percentages higher than the average are stressed in bold.

The Table 23.6 confirms the fact that in northern regions farms are more intensive and thus potentially more polluting than in the other regions. Indeed, almost all the percentages in bold refer to typologies belonging to groups G1 and G3, groups, which include farms potentially more polluting than the average within their farming type. The only exception is the one of mountains, where intensification (and pollution) in many cases gives way to farm abandonment (group NS-G4, i.e., area of deactivation and disinvolvement). On the opposite, southern regions tend to be more related to situation of higher sustainability, with the only exception of lowlands, where it is possible to find a relatively high share of potentially polluting farms, belonging to G1 groups.

The situation of central regions is in some way 'in between'. Again, in the lowland potentially polluting groups (G1 and G3) are prevailing, while hills and mountains show both potentially polluting (G1 and G3) and relatively sustainable (G2 and G4) farms.

As regards efficiency (at least from a social viewpoint), the best situation is shown in southern lowlands, while the worst refer to central lowlands and hills and to northern mountains.

However, the analysis could be improved by performing an adequate weighting procedure on RICA sample.

Table 23.6   RICA sample in 1994: farm distribution inside each typology according to geographical and altitudinal location

	NL[*]	NH	NM	CL	CH	CM	SL	SH	SM	Total
S-G1	**32.64**	**10.19**	5.40	4.86	5.71	0.93	**25.00**	9.03	6.25	100
NS-G1	**31.09**	**10.92**	6.68	**7.19**	8.40	0.51	**21.01**	9.66	4.53	100
S-G2	11.29	4.18	5.43	3.26	6.86	1.59	**19.82**	**27.09**	**20.48**	100
NS-G2	9.18	7.40	7.43	3.66	**12.20**	2.41	**18.97**	**22.66**	**16.08**	100
S-G3	**44.88**	**11.49**	3.95	3.95	2.15	0.18	15.44	12.21	5.75	100
NS-G3	30.34	**11.18**	**8.10**	**7.77**	**12.43**	1.19	14.95	9.02	5.01	100
S-G4	10.48	4.76	6.67	**5.71**	6.67	0.00	**21.90**	**26.67**	**17.14**	100
NS-G4	8.50	8.27	**9.65**	5.26	**17.19**	**3.50**	**20.17**	**18.67**	8.79	100
Average	21.08	9.21	7.53	5.68	11.49	1.76	18.67	15.27	9.30	100

[*] NL = Northern lowlands, NH = Northern hills, NM = Northern mountains, CL = Central lowlands, CH = Central hills, CM = Central mountains, SL = Southern lowlands, SH = Southern hills, SM = Southern mountains

Source: Our elaboration from RICA data bank at 1994

Conclusions

This article provides a picture of Italian Agriculture through the analysis of the RICA sample, which includes a higher share of farms with good economic and professional features than in the universe. Nevertheless, this analysis shows that part-time farms and farms of the area of disinvolvement and deactivation account for a major share of RICA farms, 30% and 18% respectively. Their situation can be considered as 'critical' in terms of actual behaviour or future trends, since part-time farms are relatively polluting and disinvolvement farms risk to result in an abandonment of lands. These farm typologies very often do not react to agricultural policies in the same way than other farms. Thus, since they account respectively for 17.75% and 20.48% of total utilized agricultural land in RICA sample (and probably more on the total Italian agriculture), policy-makers should study interventions specifically aimed to steer their management and evolution. In the case of part-time farming, that is considered important in terms of land management and conservation more than in economic and employment terms, interventions should aim to change their attitude towards the use of potentially polluting inputs, since they account for about 30% of the total amount of flows, although their share in terms of surface is much lower, only 18%. As regards farms belonging to the area of disinvolvement, their evolution towards total abandonment could be as negative as the pressure of part-time farms on natural resources. Indeed, they probably account in Italian agriculture for more than the 20% of total utilized agricultural land they account for in the RICA sample. Consequently, a negative evolution of this farm typology could bring about major problems in land conservation, once that farm activities have totally ceased.

Policy makers should be also aware of the importance of farms whose future is uncertain, since they manage as well about 20% of the total utilized agricultural land of the RICA sample. Indeed, these farms risk to evolve towards less sustainable typologies or, the more likely, to evolve towards reduced involvement and deactivation models. The degree of uncertainty in farm future performances could be highly influenced by the EU agricultural and rural policy. From this point of view, the deep changes currently going on in the European scenario could have contributed in increasing the number of farms belonging to this typology. At national level, since the main problem of these farms seems to be their small acreage, they could likely profit by a law on land renting leaving a quite high degree of flexibility to land owners and farmers. Indeed, after that law 203/1982 permitted – provided that both parties agreed on conditions – to use various types of renting contracts, almost all new renting contracts have been made using the possibility of not following the standard rules stated by law. Consequently, it seems that this freedom has helped farmers and landowners to achieve mutually satisfactory agreements on rental conditions.

As regards farm typologies that show good economic performance, the main object of decision-makers should be the one to limit their polluting potential. Indeed, farms not making an efficient use of external inputs seem to be quite limited in number and surface. From this viewpoint, it could be very important to be capable of transferring the model of large efficient farms to other farm typologies. Traditional large acreage farms, e.g., could improve their sustainability by reaching a larger physical dimension. However, almost all farm typologies could benefit by an increase in physical dimension. On the actual scenario, where the number of people making a living by agricultural activities is increasingly lowering, policies aiming to promote temporary of permanent reorganization of land use could allow achieving more suitable farm acreage for sustainability and economic viability.

Note

1   The fifteen groups relating to productive orientation are: 1) Grain cereals, 2) Rice, 3) Root crops, etc., 4) Horticulture, etc., 5) Flowers, 6) Vineyards, 7) Fruit trees, 8) Olive groves, 9) Dairy cattle, 10) Beef cattle, 11) Dairy and beef cattle, 12) Sheep breeding, 13) Mixed crops, 14) Mixed crops and herbivores, 15) Crops and livestock.

References

AA.VV. (1992), '*Strategie familiari, pluriattività e politiche agrarie*', Il Mulino, Bologna.

Andreoli, M., Colosimo, V. and Tellarini, V. (1998), 'La sostenibilità, l'efficienza ed il successo', in Pennacchi, F. (ed.), *Sostenibilità Efficienza e Successo Aziendale. Una valutazione nelle aziende della RICA*, C.N.R.-RAISA, Quaderni dell'Istituto di Economia e Politica Agraria n. 24, Centro Stampa Università degli Studi di Perugia, Perugia, pp. 71-198.

Andreoli, M. and Tellarini, V. (1995), 'Le imprese agricole e il successo', in Cannata, G. (ed.), *Per una definizione delle condizioni di «successo» nelle aree collinari e montane*, C.N.R.-RAISA, Rubbettino, Catanzaro.

Barnard, C.S. and Nix, J.S. (1973), '*Farm planning and control*', Cambridge University Press, London.

Cannata, G. (ed.) (1995), '*Per una definizione delle condizioni di «successo» nelle aree collinari e montane*', C.N.R.-RAISA, Rubbettino, Catanzaro.

De Benedictis, M. (ed.) (1995), 'Agricoltura familiare in transizione', INEA, Roma.

Pennacchi, F. (ed.) (1994), '*La riforma Mac Sharry. Effetti nelle aziende RICA*', C.N.R.-RAISA, Quaderni dell'Istituto di Economia e Politica Agraria, Centro Stampa Università degli Studi di Perugia, Perugia.

Pennacchi, F. (ed.) (1998), '*Sostenibilità Efficienza e Successo Aziendale. Una valutazione nelle aziende della RICA*', C.N.R.-RAISA, Quaderni dell'Istituto di Economia e Politica Agraria n. 24, Centro Stampa Università degli Studi di Perugia, Perugia.

PART 4
EDUCATION AND SUSTAINABLE DEVELOPMENT

24 Higher education as a means to sustainable development in marginal regions

LENNART ANDERSSON AND THOMAS BLOM

Background and purpose

Institutions of higher education in themselves and an increased participation in higher education among the inhabitants are, directly and indirectly, seen as important means of alleviating the negative implications of marginality. Higher education institutions are important producers of scientific knowledge. An increase in the knowledge level of the inhabitants and the diffusion of scientific knowledge to those working in agriculture, manufacturing and services are, without doubt, major assumptions in the work of creating development and sustainability (Andersson and Blom eds., 1998). There are, however, few geographical studies in this research field.

Society today is very often organized in sectors. Plans and intentions are worked out for different sectors and decision levels. Often there are special plans for the global, national and local levels. The plans for different levels are normally connected with each other, at least in a formal way. Too often, the idea is that the intentions for the sector will flow from the global via the national to the local level where those people are living who are the ultimate target group. In reality, the complexity of life and society disturbs the implementation process. Experiences from reality and research tell us that reality often is a pale copy of the intentions (Andersson, 1998). There are many reasons for penetrating the relations between global intentions, national and local goals and the local reality in marginal areas as regards issues of higher education and the diffusion of scientific knowledge. This is the background to our paper.

Geographical variations in participation in higher education

Higher education is facing a major expansion, which will continue under the combined effect of two factors: population growth (particularly in the developing and least advanced countries) and the demand for increasingly higher levels of education. In Table 24.1 we can see gross participation rates in higher education in 1985 and 1995 in different regions of the world.

Table 24.1   Gross participation rate: higher education students, regardless of age, as a percentage of the population of the 5-year age group following the end of secondary schooling

Regions	Enrolments	
	1985	1995
WORLD TOTAL	12.9 %	16.2 %
Developed regions	39.3 %	59.6 %
North America	61.2 %	84.0 %
Asia / Oceania	28.1 %	45.3 %
Europe	26.9 %	47,8 %
Countries in transition	36.5 %	34.2 %
Developing regions	6,5 %	8,8 %
Sub-Saharan Africa	2.2 %	3.5 %
Arab States	10.7 %	12.5 %
Latin America / Caribbean	15.8 %	17.3 %
East Asia / Oceania	5.4 %	8.9 %
incl. China	2.9 %	5.3 %
South Asia	5.3 %	6.5 %
incl. India	6.0 %	6.4 %
Least advanced countries	2.5 %	3.2 %

Source: UNESCO (1998), Original source: World Education Report 1995

There has been an increase world-wide in higher education participation rates between 1985 and 1995. In the developed regions, participation rates rose considerably between 1985 and 1995 (from 39.3% to 59.6%). This trend will continue, despite an ageing population, as the demand for higher educational levels is very strong. In the developing regions, higher-education participation rates are markedly lower than in the developed countries (in 1985, 6.5% vs. 39.3%; in 1995, 8.8% vs. 59.6%). The gap has widened in 10 years, the increase in participation rates in the developing regions being less marked than in the developed regions. In the developing regions, the sub-Saharan African region gives cause for particular concern because it has the lowest participation rates (2.2% in 1985 compared with 3.5% in 1995). Its situation is fairly similar to that of the least advanced countries (2.5% in 1985 compared with 3.2% in 1995).

Turning from the global level to a comparison of conditions in a number of individual countries, we find considerable differences. Figure 24.1 shows the percentage of the population (aged 25-64) with at least upper secondary education and the percentage with some post-secondary or higher education for a number of OECD countries. It should be noted that the choice of countries has been limited by the availability of current statistics.

Canada has the highest percentage with post-secondary education and is followed by the US, Norway and Sweden, whilst Spain, Poland and Portugal have lower values. Whereas 46.9% of the population of Canada aged 25-64 have post-secondary education; the equivalent figure for Portugal is 11%.

Within individual countries, there are similar or even greater differences between regions and municipalities. For instance, the highest values for Sweden are to be found in metropolitan regions and municipalities with universities. The lowest proportion is to be found in marginal regions and old industrial areas (SCB, 1997).

Even in metropolitan areas in Sweden, there are considerable differences in the level of education. In the municipality of Danderyd, which is part of the Stockholm metropolitan region, 37% of the population between the ages of 25-64 in 1997 had completed three or more years' post-secondary education. The figure for the municipality of Botkyrka, which is also part of the Stockholm metropolitan region, was 9%. A low level of education is also a problem in metropolitan regions (SCB, 1997).

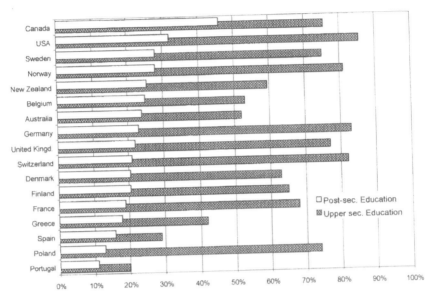

Figure 24.1   Level of education, percent of the population (aged 25-64) with at least upper secondary education and population with post-secondary education for some OECD countries

Source: OECD (1997)

Global, national and local visions and goals for higher education

Global visions and action

As a basis for global visions and action for higher education we refer to and discuss the working document named 'Higher education in the twenty-first century. Vision and action' which was drawn up for the world conference of UNESCO in Paris in October 1998 (UNESCO, 1998). The document is chiefly based on the documents and final declarations of five regional conferences, held by UNESCO in preparation for the world conference. The document has been revised by the Executive Secretariat of the Steering Committee of the UNESCO Consultative Group for Higher

Education, which was responsible for overseeing the preparations for the world conference in 1998.

UNESCO's vision may be described as centred on ten principles. We indicate below those that are of special interest for the purpose of this paper:

1) The universality of higher education presupposes universal access for all those who possess the requisite abilities, motivation and suitable preparation, at all stages of life.
2) The universality of higher education presupposes the employment of varied forms of intervention to meet educational needs for all and at all stages of life.
3) The universality of higher education implies that it should build up links of universal solidarity with other institutions of higher education and with other institutions of society.

The document also states that the contextualized and universal nature of higher education makes it easier to relate thinking about it to four pivotal issues, namely, pertinence, quality, management and funding, and co-operation. In the context of this paper, the plans for a new policy of co-operation are of special interest. In response to the changes affecting society and higher education at the dawn of the twenty-first century, a new policy needs to be devised. Some points may be mentioned here:

1) Taking account of the phenomena of regionalization and globalization, and of the fact that society is increasingly becoming a knowledge-based society, co-operation policies must keep both abreast and ahead of these developments by making teaching and research programmes more international, but at the same time establish mechanisms to combat polarization, marginalization and fragmentation.
2) Taking account of the arrival upon the scene of IT, co-operation policies must enable the developing countries to equip themselves with the means of access to it.
3) In order to put into reverse the process of decline affecting institutions in the developing countries, the least developed in particular, co-operation policies should aim for greater independence and closer partnership between the world of work and higher education.
4) Policies for stepping up co-operation with other institutions of higher education and other levels of education should be implemented, so that education becomes an unbroken chain.

5) The institutions of higher education in developing countries may be looked upon as being essential to the overall development of the education system and to the promotion of endogenous capacities.

6) Strengthening the key components in the integration and co-ordination of sectors, institutions, agencies and resources, in order to ensure better horizontal co-operation.

7) Defining a relevant policy and an overall plan of action jointly with the various partners involved, on the basis of an analysis of past or ongoing experience and of anticipation of their knock-on effects and the new priority challenges to be met.

8) Helping managers in defining, implementing and following up their co-operation policies and in pooling their experiences.

In another section of the document, the point is made that we are living in a period that is ushering in profound and irreversible changes, but also a period when decisions are called for. In order to take sound decisions, it is necessary to analyze and bring into the open the paradoxes that come with the decisions. Some of them are listed below:

1) Alongside a powerful move towards globalization of the economy, there exists a trend towards the creation of more small and medium-sized enterprises, or indeed towards an informal economy.

2) Firms are increasingly relocating to places where a labour force with low skills is available at very low cost, while highly skilled personnel are more and more required to move to where goods are produced and decisions taken.

3) A growing demand for higher education in developing countries which states have difficulty in meeting at the same time as a lengthening of the time spent in education in rich countries, in order, among other things, to reduce unemployment.

4) A widening gap between developing and rich countries concerns their capacity to provide themselves with the basic infrastructure for access to the exponential growth of scientific knowledge (including information technology).

As its final point, the document states that when confronting the paradoxes it is important to transcend the contradictions, i.e. not to reason in terms of 'either/or' but in terms of "both the one and the other, depending on the context", since both vision and action need to be situated. But, behind this plurality, however, lie certain basic needs common to all humanity. The final statement is, consequently, that, if vision and action are to be situated, they should be inspired, first and foremost, by a universal vision, a vision of the organizing of a fairer and more equitable society.

To summarize, the UNESCO document on vision and action in higher education in the twenty-first century has brought to the surface some essential problems with the diffusion of higher education, which are also valid for the work on development and sustainability in marginal regions.

National goals

In Sweden one of the goals of higher education policy, as well as of regional policy, is to widen the opportunities of participating in higher education for individuals living in those parts of the country that are influenced by marginality. New higher education institutions have been established in parts of the large cities suffering from marginality and in areas with a low population density and other characteristics of traditionally marginal areas. Many measures are directed from the national level but presuppose functioning local co-operation between universities and the local communities. In the regional policy measures adopted by the parliament and the government, promoting the diffusion of higher education and scientific knowledge to neglected areas and groups of people plays an important role. The establishment of higher education institutions as a means in regional policy is also valid for other countries, e.g. Norway, Finland, Great Britain, The Netherlands and Italy (Cederlund, 1999).

Local goals

In local visions and development programmes for marginal areas the issue of how to work together with universities and how to increase the degree of participation in higher education among the inhabitants always plays a very important role. The local politicians have a very strong belief in the positive effects of higher education in the work for better development. They are inhabitants in the community but also the link between the inhabitants and the intentions and plans stemming from the regional and national decision-makers. There is, however (Blom, 1996, p. 199) a feeling that the world of politicians, even at the local level, has 'bought' the centrally initiated perspective and tends to forget the views of the inhabitants. Kåpe (1999) reports the same view for the politicians connected with medium-sized Swedish cities, where most of the new universities are situated.

If we consider the general goals that local politicians in marginal communities in Sweden tend to stress, the primary one is increasing the proportion of inhabitants with some form of higher education. At the same time, there is awareness that there is a rather restricted demand for higher educated people. There is also an awareness that many people with higher education are moving away from marginal areas. Not many of them are coming back again.

Local reality in marginal areas

The goals stated by decision-makers on different decision-levels concerning higher education may in general be regarded as something different from local reality, that is the importance of higher education in the real lives of individuals and in the real development work for a sustainable society. The presence of 'unintentional implications' plays tricks (e.g. Andersson, 1998). We mentioned above that UNESCO in its working document has stressed the fact that higher education may be looked upon from a situated point of view, that is the impact of higher education differs between different environments. Geographers have described similar pictures of local reality. We shall refer to two Swedish studies made respectively by Hillbur (1998) in three villages in north-western Tanzania and by Blom (1996) in three marginal communities in western Sweden. A third Swedish study, by Kåpe (1999) will be considered as well. Kåpe has evaluated the real implications of the extension of higher education and research in medium-sized Swedish cities. People's view of higher education and scientific knowledge in an Indian village will also be discussed.

Diffusion of agricultural knowledge in three villages in Tanzania

Hillbur (1998) analyses the diffusion of agricultural knowledge to achieve better planning for the sustainable use of natural resources at the local level. He distinguishes between two knowledge profiles. One of these is termed the system perspective or top-down perspective, which is dominated by sectoral specialization and is represented by national policies, donor agencies and scientific specialization. This profile may be said to be the perspective which stresses scientific knowledge and knowledge gained from higher education. He maintains that the system perspective is guided by the principle of similarity. The other perspective is a knowledge profile with a territorial delimitation, which he terms the arena perspective. This profile reflects principles of nearness and the bottom-up perspective. The arena perspective, which is shaped by the way, individual human beings organize their daily lives, may often be said to be 'the layman's perspective'. It would correspond to local institutions, farmers' activities and their knowledge of the environment. Hillbur argues that "we all live in a context which is influenced by our understanding of local processes and phenomena, even though we, in our roles as professionals, often suppress this in favour of sectoral and specialized competence." The two perspectives should be seen as complementary. However, it is common for them to be separated, which leads to barriers for communication and a distinction between the public/professional sphere and the private sphere (Hillbur, 1998, p. 168).

The result of his fieldwork indicates that approaches to agricultural extension would benefit from a perspective that is sensitive to nearness as regards social organisation. The purpose of the knowledge arena, i.e. to combine neighbourhood (nearness) or location-specific factors with local processes moving across landscape units has a practical applicability. Implementation of complex issues in the agricultural sector must therefore be carried out in association with a socially acceptable definition of agricultural development (Hillbur, 1998, p. 173).

The village Chamaon in India, higher education and scientific knowledge

The following section is a report from an interview with two scientists, Gerhard Gustafsson and Kristina Lejonhud, who, during the last decade, have been conducting research on changes in the Indian village Chamaon, seven kilometres outside the city of Varanasi in India. The actual village has about 2400 inhabitants in 1999. Its economic life is still centred on agriculture but with a growing element of handicraft and manufacturing and of commuting to work places mostly in the service sector in Varanasi.

Participation in elementary education has increased considerably during the 1990s. A state owned elementary school and private schools are now established in the village. After elementary school, the pupils need some years of college education in Varanasi to be qualified to start university studies. The interest in higher education is caste-related. Parents belonging to the higher castes want their children to go on to higher education. They are also concerned with quality and some send their children to elementary schools outside the village. In occasional cases, even people from the lower castes are interested in higher education for their children.

People from lower castes have priority when it comes to the admission of new university students in India and, in the end; this may probably lead to reduced inequalities. However, people do not talk very much about education. Other things in everyday life predominate. There are less than five people with higher education living and working in the village. Those with higher education who come from the village live and work in white-collar professions in Varanasi. The fifteen teachers at the elementary school in the village all commute from other places in the Varanasi region.

The most evident economic developments during the 1990s have taken place in agriculture. The local knowledge on how to manage agriculture is formed in different ways. One major way is observing one's neighbours and then applying the same methods etc. There are examples where an well-educated farmer has been able to obtain new knowledge outside the village. He has made use of his new knowledge and it has gradually spread among his neighbours. Important are also the local public advisers on agriculture, who have probably been indirectly linked with scientific knowledge. New knowledge has influenced agriculture as regards the preparation of land, seed and insecticides. However, it is still not possible to introduce new methods for harvest because farming is done on small strips of land.

A concluding reflection may be that links with higher education and science are not, at present, considered to be important means of developing the socially complicated and traditional life in the Indian village in question.

Attitudes to higher education in three Swedish marginal communities

In his investigations of attitudes to higher education in three Swedish marginal communities, Blom (1996) presents in principle the same basic ideas as Hillbur, though there is no other relationship between the two studies. Blom contrasts what he terms the everyday knowledge of the individuals in a local community and formal knowledge, by which he means knowledge gained in the formal school system, in our case the higher education system. These two approaches resemble the concepts used by Hillbur, the knowledge-arena perspective and the system perspective. The low participation rate, for Swedish circumstances, that Blom finds in the three communities is often considered as a problem in the public debate. It is important, however, Blom argues, that this situation is seen as a stimulus to make a closer analysis of ways to utilize the everyday knowledge potential of the territory and create an interplay with the external opportunities provided by the higher education system. It may also be looked upon as a request to the higher education system to develop education that may foster co-operation between everyday knowledge and formal systematic knowledge and provide better support for a relevant local process of development.

There are obvious differences between the three communities but Blom claims that the "ordinary" inhabitants look at higher education from a typical insider perspective. Higher education is neither possible nor meaningful to force them to accept. They stress the importance of everyday knowledge.

The typical industrial entrepreneur in these marginal territories maintains that there is little need of people with higher education if you consider the issue from the situation today. At the same time he stresses the importance of improving the competence of the staff by means of different forms of mostly internally organized further education.

It is often said that the structure of marginal communities is well integrated and that everybody knows everybody. Blom thinks that this is often not the case. The different actors meet each other in often very segmented roles. It is necessary to create a more integrated local network with more links to the regional and national levels and to the sectoral institutions, in our case the universities.

The university and the medium-sized Swedish city

We should also like to refer to another study, not dealing with marginality but nevertheless of interest. In his study, which considers the medium-sized Swedish cities from a population perspective, Kåpe (1999) has analyzed the impacts of the universities located in these cities. The universities have proved to be the most important public institution. They have replaced military garrisons as the symbol for the status as a major medium-sized Swedish city. Kåpe's findings may be related to Blom and the frame used by Hillbur. Kåpe refers to other studies, which indicate that the real impacts of the universities on the business life of these cities often tend to be overestimated. Kåpe does not explicitly draw this conclusion but the essence is that the university sector has not succeeded in integrating its activities with the activities and existing competence of the territory and its economic life. Another interesting observation is the relatively little interest from the policymakers in stressing the most important task of the university, namely increasing the general educational level of the inhabitants. Here the universities in the medium-sized cities of Sweden have been fairly successful if we take into account the city itself and its closest surroundings. However, there are many indications that the geographical reach of the university is small. The influence of the university on the marginal communities which are often situated on the outskirts of the regions of these cities is small, as is also documented in the study by Blom mentioned above.

The studies we have considered point to the need of a thorough penetration of the issue of integrating the sectoral institution of the university with the territorial community with its existing everyday knowledge and existing local competence for producing certain kinds of goods and services.

How can new knowledge be formed in marginal regions?

Local and regional restrictions

The way in which knowledge is formed that can foster development and sustainability in marginal areas is seldom dealt with. It may be said that new knowledge is formed anywhere two or more people meet. In that process, the proximity factor plays an important role. In some sense, all knowledge is regional. The role of the university is to create the necessary conditions for an increase in territorially generated knowledge. One condition for an increase is a certain amount of confidence between the university and its local partners, and between the different groups within the territory. The feeling of uncertainty between the interacting partners must be reduced. In reality, there are many barriers.

Hägerstrand (1975) maintains that there always are some regional restrictions which people have to submit to. In different regions, different rules and norms predominate. Different groups or individuals have different rights and obligations. Social heritage is one of the restrictions. The possibility of choice is also restricted by the demands and wishes which the social environment places on individuals (Persson 1999). In Sweden and some other countries with remnants of an old industrial structure in many small places in marginal areas, the heritage of dependence on the industrial enterprise even for social life is important and still alive. This phenomenon reduces the scope for action that is available to the individual and is generally thought to have conserving implications.

During the last few years, great efforts have been made in Sweden to increase the level of knowledge in marginal regions. According to the higher education ordinance adopted by the parliament in 1997, universities have three main tasks: to provide higher education and to conduct research and, thirdly to co-operate with the surrounding society. The third task was added in a situation where the existing co-operation was considered unsatisfactory. Two examples are given in the following section of the paper. One reports some experiences from research on the mechanisms etc., which are important for successful co-operation between university and society. The other describes a project whose purpose is to introduce university activities into communities in the county of Värmland.

Increasing research on 'universities' third 'task'

We have studied the abstracts of papers presented at a conference on research on the third task organized by the Dalarna University in Sweden in March 1999 (Högskolan Dalarna, 1999). About 50 papers dealing with different projects and case studies were presented. We summarize below the issues and conclusions raised in the conference, which we feel are relevant to our paper and in so doing, concentrate on the process whereby knowledge is developed in the local territory.

In order to understand the role of a university, it is necessary to consider the characteristics of the regional and national environment as well as the characteristics of the university. The prerequisites for research-practitioner collaboration concerns first of all the ways in which we relate to each other and secondly the development of a common language. It is essential to analyze the social production of knowledge, in which such concepts as ethics, humanism, democracy, pragmatism, must be used in connection with science, technology and economy.

The use of knowledge is conceived as a process of social interaction and informal learning. Contrary to the traditional separation between the generation and use of knowledge, new knowledge is assumed, to a certain extent, to be generated in practice as ideas are interpreted and used in new contexts. It is important to inquire into the epistemology of action knowledge, i.e., into theories, strategies and methods governing people's actions in social practices. The focus must shift from the organisation of the university itself to the very process of organizing the activities it undertakes.

What is central here is to take into account the significance of the dialogue between the local community and science and the co-construction of mutual relevant knowledge in what can be termed the local arena. This dialogue concerns the production of knowledge and what can be accepted as knowledge. It would seem to be of importance that certain types of knowledge are not excluded and declared invalid from the beginning by reference to some implicit academic model of scientific knowledge. The focus of the learning process is a combination of inside-out and outside-in perspectives. Through collaboration with practitioners as equal partners academia obtains access to knowledge and processes of inquiry inherent in praxis, as well as an opportunity to play an active role in local processes of learning, reflection and development. The theoretician must, to a greater extent than now, take into consideration the ways in which working men interact. Research circles, with scientists and people from the local territory as participants, are looked upon as a valuable instrument for acquiring new knowledge on development issues. Even other types of action-oriented research are favoured.

Generally, it may be said that the sector perspective, the university perspective, predominates in the papers we have referred to, which is a weakness. It may also be noted that the doubtful attitudes of the inhabitants of, for instance, marginal areas in Sweden towards higher education and science are not taken into account.

Measures to increase the level of education in the county of Värmland

In Table 24.2 we note the measures taken by the different municipalities and the proportion of the population who were new university students in 1989, before the project started and in 1993 while the project was in progress.

During the period 1992-1995 there was a project in the county of Värmland, Sweden whose objective was to increase the level of participation in university studies among the people in the different communities. The background was that the county of Värmland, the university town of Karlstad and the municipalities surrounding Karlstad excluded, had a level of education that was below the average for the country. The project was based on a programme of action where every community had worked out its own strategy. The physical distance to the university was bridged with

Table 24.2   Measures to increase the number of new university students in the municipalities in Värmland

Municipality	Take-up rate for higher education (per 1000 inhab. 1989/90)	Co-operat. with Karlstad University	Municipal and industrial joint work groups on educat.	Co-ordin. of education needs	Request for de-centralised courses offered	Study centres with video conference equipment	Distance courses from university by eletronic means	First class comm. with university	Take-up rate for high educ. (per 1000 inhab. 1993/94)
Arvika	8.4	yes	yes	yes	1/5	yes	3	yes	11.4
Eda	4.9	yes	yes	no	3/1	planned	2	no	9.5
Filipstad	5.7	yes	yes	yes	8/2	yes	3	yes	6.4
Forshaga	*7.5*	*not incl.*							*12.5*
Grums	5.9	yes	no	no	6/4	no	2	yes	9.9
Hagfors	6.5	yes	no	yes	3/1	yes	4	no	6.8
Hammarö	*7.8*	*not incl.*							*12.0*
Karlstad	*10.0*	*not incl.*							*13.8*
Kil	*9.1*	*not incl.*							*11.3*
Kristinehamn	7.9	yes	no	no	7/1	yes	3	yes	10.6
Munkfors	2.6	yes'	no	yes	8/0	yes	2	yes	13.2
Storfors	6.4	yes	yes	yes	5/1	planned	1	no	8.4
Sunne	5.8	yes	no	yes	5/5	yes	2	yes	10.3
Säffle	7.5	yes	yes	yes	3/3	yes	3	yes	10.3
Torsby	7.0	yes	yes	yes	14/1	yes	5	planned	7.2
Årjäng	4.5	yes	yes	yes	7/3	yes	3	yes	5.2

Source: Lindqvist, 1995

the aid of electronic communication. The mental distance between the university and the municipality and between different actors in the municipality was counteracted by the formation of different types of groups, associations, meetings and communicative activities. The county board supported the municipalities financially.

The participating municipalities all suffer, in different degrees, from the symptoms of marginality. The municipalities of Eda, Torsby and Årjäng situated 100-150 kilometres from Karlstad.

For most of the municipalities, there was a considerable increase in the proportion of new university students, young and adults, during the study period. It is not possible, however, to decide how much of this increase was conditioned by the project. There was a considerable increase for the country as a whole during the period in question. For some of the marginal municipalities, the increase has meant a change, from both a national and regional point of view, from a low level to a level that corresponds to the average.

The picture reflected by Table 24.2 and Figure 24.2 shows that even in a state like Sweden there are major differences between the different municipalities. It is logical to say that the sectoral institution, the university, is the same, but that the territories are unique and must be dealt with on the basis of their own original conditions. Reality in Värmland confirms the statement made in the UNESCO working document referred to earlier in this paper, that the measures taken by universities must be linked to the character and identity of every region and community. The problems of the interplay between the sectoral institution and the territory must therefore be dealt with on the basis of this statement. This observation also confirms the basic ideas in the geographical studies discussed earlier.

The amount of co-operation in the municipalities concerning higher education issues differs as does the business climate (Karlsson, *et al,*. 1999). The establishment of co-operative groups for higher education between municipal representatives, entrepreneurs and other groups are consequently an important part of the project. It can be seen from Table 24.2 that such co-operative groups have been established in most of the municipalities, but not in all of them. These groups work on such issues as information, the changing of attitudes, the co-ordination of the demand for education and Cupertino with the university.

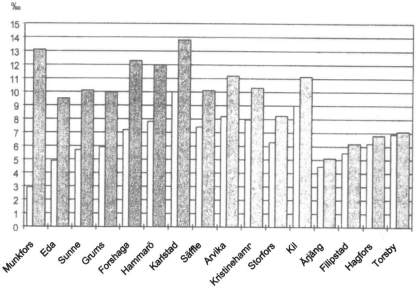

Figure 24.2   The proportion of new university students in the municipalities in Värmland, Sweden, in 1989/90 and 1993/94, per thousand inhabitants

Source: Lindqvist, 1995

Karlstad University has more recently established special university offices in some places in the county with the aim of overcoming the distance factor.

As a result of the work in the project, different types of university courses are being increasingly organized with the aid of computers and special videoconference equipment and discussion groups using first class communication. Special subsidies are provided for the establishment of special learning centres, where the computer equipment is located.

A principally important issue is which type of education and scientific knowledge is the best to distribute locally with the aim of fostering development in the specific territory – education of a general kind or education with a close link to the specific enterprises established in the municipality. We believe the best is a locally adapted, mixture of regionally specific knowledge and more general knowledge. This makes it possible for the municipality to develop its own specialized competence and also to integrate knowledge for general renewal.

Some concluding remarks

Higher education is an important means of fostering sustainable development in marginal regions. However, there are evident difficulties in evaluating the implications of different measures. These may only be visible in a long-term perspective. One reason for this is that these implications represent only one of many factors influencing development and sustainability in marginal areas. It is therefore of importance to discover whether there are any specific characteristics in the diffusion process, which need to be taken into account.

Some important observations may be noted:

1) Even the global level document, the UNESCO document, notes the geographical factors that affect the diffusion process and cause differences and paradoxes.
2) The most striking element is the enormous differences in participation in higher education between different parts of the world. The most important issue is to analyze how higher education and scientific knowledge may be linked to regions and municipalities in those areas with low participation rates.
3) There is an obvious difference in local territories between the attitudes of politicians with their great appreciation of the value of higher education and the uncertain and often negative opinion among the inhabitants and even among the representatives of the economic life of the municipality.
4) In the geographical studies, the local reality has been described against a background of the relations between the characteristics of the sector, the university, and the other dimension, that is the characteristics of the life in the territory in question. One way to characterize these two dimensions is to stress the relations between the systemic diffusion model of scientific knowledge and the everyday knowledge that is developed in the knowledge arena, that is the local territory. Reality suggests that the knowledge developed in the local territory is important and a necessary basis for cooperation between the university and the territory. The proximity factor plays an important role. In some sense, all knowledge is regional or local. Knowledge is generated wherever two people meet.

5) The role of the university is to create the necessary conditions for an increase in territorially generated knowledge. One condition for an increase is a certain amount of confidence between the university and its local partners, and between the different groups inside the territory. The sense of uncertainty between the interacting partners must be reduced. In reality, there are many barriers.

6) Computers are surely an effective means making higher education important in the development process. However, this equipment is expensive and this is a barrier in many parts of the world.

References

Andersson, L. (1999), 'On the Issue of Unintentional Regional and Local Impacts', in Gourlay, D. (ed.), *Marginal Rural Areas in the New Millennium. New Issues? New Opportunities?*, The PIMA98. International Conference at The Scottish Agricultural College, Aberdeen, Scotland June 26 - 29 1998.

Andersson, L. and Blom, T. (eds) (1998), 'Sustainability and Development. On the Future of Small Society in a Dynamic Economy', *Research Report* 98:8, Social Sciences, University of Karlstad, Sweden.

Blom, T. (1996), 'Perspektiv på kunskap och utveckling. Om attityder till högskoleutbildning i några perifera regioner' (Perspectives on Knowledge and Development. On Attitudes to Higher Education in Some Peripheral Regions), *Research Reports Social Sciences*, 1996:3, University of Karlstad, Sweden.

Cederlund, K. (1999), 'Åter till universiteten' (Back to universities), in Geografi i Lund, Essäer tillägnade Gunnar Törnqvist, Lund University, Sweden.

Hillbur, P. (1998), 'The Knowledge Arena, Approaching agroforestry and competing knowledge systems - a challenge for agricultural extension', *Meddelanden från Lunds universitets geografiska institutioner, Avhandlingar nr* 136, Lund University Press, Lund, Sweden.

Hägerstrand, T. (1975), 'Survival and Area. On the lifehistory of individuals in relation to their geographical environment', *Monadnock*, Vol. 49, Clark.

Högskolan Dalarna (1999), *Högskolor och samhälle i samverkan* (Universities and Society in Cooperation), Sammanfattning av papers presenterade vid HSS-99, Dalarna University, Sweden.

Karlsson, S., Stensmar, M., and Ednarsson, M. (1999), *Visst blåser det olika. En studie av näringslivsklimat i fyra värmländska kommuner* (Of course it blows differently. A study of the climate of economic life in four communes in Värmland), Arbetsrapport, February 1999, Karlstad university, Sweden.

Kåpe, L. (1999), *Medelstora svenska städer, En studie av befolkning, näringsliv och ortssystem* (Medium-sized Swedish Cities, A Study of Population, Economic Life and System of Places), Karlstad University Studies 1999:2, Karlstad University, Karlstad. Sweden.

Lindqvist, M. (1995), 'Värmlandsprojektet för att öka rekryteringen till högskolor och universitet' (The Värmland project to increase the recruitment to universities), Slutrapport, PM, Högskolan i Karlstad, Karlstad, Sweden.

OECD (1997), *Education at a Glance* – OECD Indicators 1997, Paris.

OECD (1998), *Education at a Glance* – OECD Indicators 1998, Paris.

Persson, K. (1999), 'Makt och lokal utveckling, Vem och vad styr politiska beslut för rummets utformning' (Power and Local Development, Who and what controls political decisions and the design of space, *Karlstad University Studies* 1999:4, Karlstad University, Karlstad, Sweden.

SCB (1997), *Befolkningens utbildning*, Statistiska meddelanden, U37SM9701.

UNESCO (1998), *Higher Education in the Twenty-first Century, Vision and Action*, World Conference on Higher Education, Paris 5-9 October 1998, Working Document,

UNESCO (1999), *Statistical material from UNESCO homepage*, January 1999.

Other sources

Interview (1999-04-19), with Gerhard Gustafsson and Kristina Lejonhud, Dept. of Geography and Tourism, Karlstad University, Sweden.

25 The development and implementation of sustainable education – northern areas, Pakistan

LUCY JONES

Introduction

Pakistan covers an area of 887,700 square kilometres, which is approximately a third of the size of India and almost four times as big as the UK. It has a population of approximately 140 million. Bounded by Iran to the south-west, Afghanistan to the west and north, China to the north-east, India to the east and the Arabian Sea to the south, (see Figure 25.1), Pakistan has existed as a political unit since the partition of British India in 1947 when it was created as a 'homeland' for India's Muslims. Since its creation it has had a difficult time, development being hampered by a medieval agricultural system, widespread illiteracy, a weak economy and a civil service afflicted with corruption and lethargy. The foreign policy of Pakistan is dominated by a fear of India and this paranoia has resulted in the creation of a military class heavily involved in politics. Pakistan has spent most of its short life under martial law.

Topographically, Pakistan can be divided into six regions as shown in Figure 25.2. This paper focuses on the region known as the Northern Areas, the meeting place for three major mountain ranges - the Hindu Kush, the Himalaya and the Karakorum. The elevation and extremes of climate dominate day to day life. Routes into the region are restricted and all experience frequent disruption from snowfall, flooding and in particular, from avalanches.

Prior to 1947, Pakistan was exploited as a 'colony' to provide a source of raw materials for the rest of India. This practice resulted in it having very little of its own indigenous industry. Pakistan's growth of approximately 5.7% p.a. over the past decade is therefore highly commendable and

335

is regarded as being amongst the highest in the developing nations. The economy is dominated by the service sector, which contributes an estimated 47% of G.D.P. with agriculture contributing 36% and manufacturing, 17%. This economic growth, however, is hampered by a population growth rate of 2.67% p.a. It is estimated that 50% of the population (70 million) is in the age range 0 to 18 years.

Figure 25.1   Location of Pakistan and its neighbours

The educational background

Only a half of Pakistan's school-age children attend school and the literacy rate is estimated at only 26%. Of the Pakistani children who begin school, between 70 and 80% complete Class 5 and of the children who complete the eight years of the primary cycle, an estimated 30% have attained only basic literacy levels. Disparities between male and female education are severe; only 52 girls attend school in Pakistan for every 100 boys compared with 81 in Bangladesh and 74 in India.

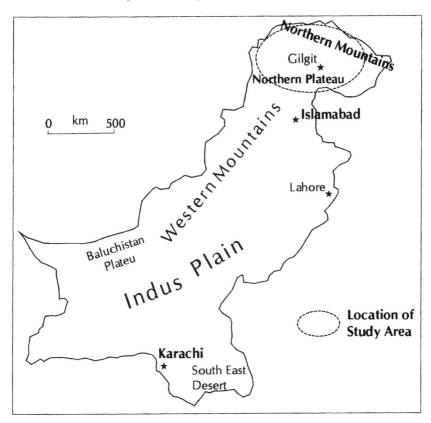

Figure 25.2   Main physical regions of Pakistan

The Government of Pakistan (GoP) has stated that it is committed to a policy of universal primary education and to the eradication of illiteracy and innumeracy in both rural and urban areas. Since 1955, the GoP has launched eight Five-Year-Plans to reform primary education throughout the country. Table 25.1, lists these Five-year-plans, their declared objectives and summarizes the reality of the situation. Within Pakistan, the Northern Areas (NA) occupy an area of 83,800 square kilometres, the size of the Irish Republic. It consists of five districts shown in Figure 25.3. The Northern Areas is not treated like a province but as a division, answerable directly to the federal Minister of State for NA in Islamabad. The people have no

Table 25.1 Pakistan, five-year-plans 1995-1998: objectives and reality

Plans	Objectives	No. of illiterate (000s) Total	l
1st Plan (1955-60)	*'the country may reasonably hope to achieve a universal system of free and compulsory primary education by about 1975'*	20,975	11.
2nd Plan (1960-65)	*'girls will be provided with much greater facilities for education and this will be done by admitting more girls to the existing schools'*	23,731	12.
3rd Plan (1965-70)	*'the objective of the third plan is to greatly increase enrolment at the primary level in order that universal primary education may be achieved'*	26,721	14.
Non-plan period	*'the aim is to create a literate population and an educated electorate by mobilising the nation and its resources'*	32,811	17
5th Plan (1978-83)	*'the plan will provide 100% coverage to 5 year old boys in class 1, so as to lay the foundation of universal enrolment in 1987'*	37,269	20
6th Plan (1983-88)	*'serious effort will be made to institute universal education by ensuring that all boys and girls of the relevant age group get enrolled in class 1 by 1988'*	42,372	23
7th Plan (1988-93)	*'the seventh plan will provide primary education facilities to all children in the age of five to nine years'*	49,000	28
8th Plan (1993-98)	*'the eighth plan will provide primary education facilities at a reachable distance for every boy and girls of the relevant age'*	50,827	29

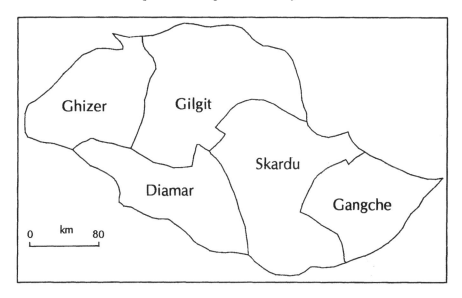

Figure 25.3   Administrative districts of Northern Areas

democratic power to elect representatives to the federal parliament and cannot take cases to the Supreme Court.

Education in the Northern Areas

The NA is among the most least-developed areas of Pakistan in terms of educational indicators. The literacy rate of 14.7% (1981 Census), 24% for males and 3% for females was well below the national average of 26%. Enrolment at the primary level covers just over half of school-age children (54.5%), with 74% of boys and 35% of girls being enrolled. These statistics mask significant inequalities between the five administrative districts. For example, in the Diamer district, girl enrolment at the primary level is only about 14% of the corresponding school age.

The Government is the largest provider of educational services in the NA with direct responsibility for 802 schools (primary, middle, and high schools) and about 84% of enrolments. School types include Mosque Schools, army public schools, non-formal Government Schools, Social Action Primary Schools, Community schools, coaching centres and private schools. The Aga Khan Education Service (AKES) is the second largest

Table 25.2   A structural analysis of major types of schools in the Northern Areas

Institution providing Buildings/Teacher salary	Types of School	Additional Information
Federal Government Schools		
GoP / GoP	Kachi, Primary School (5 - 8 years) Kachi, Middle School (8 - 10 years) Kachi, High School (10 + years)	- at primary level the medium of teaching is Urdu, English as a foreign language starting at class 6 - fees are minimal (Rs 2 - 7) - often only one teacher in primary school - lack of facilities for girls in middle and higher education
Mosque Schools		
GoP, donations, funds from religious organisations / GoP, Private	Kachi Primary Schools (5 - 8 years)	- GoP pays directly and in some cases indirectly for the teacher - in Skardu and Ganche 'private' schools with money from Kuwait and Saudi Arabia (up to middle or secondary level)
Army Public Schools		
Army / Indirectly GoP / Army	Kachi, Primary School (5 - 8 years) Kachi, Middle School (8 - 10 years) Kachi, High School (10 + years)	- fees for children of army members are approx. Rs. 50, much higher for non-army members - have a very good reputation - financed indirectly through government and the high fees
Non-Formal Government Schools		
GoP / Community	Primary School (3 - 8 years)	- 3 year education - students provided with minimum primary education in Urdu, Maths and Islamyiat - no fees charged - multi-grade teaching - many are shelterless
SAP Schools		
GoP under SAP II World Bank / Community	Kachi, Primary School (5 - 8 years)	- founded within the framework of the Social Action Programme - teachers often supported by volunteers - under NEP around 500 communities will get a new building on the basis of contributions by the GoP and communities

Table 25.2   continues...

Institution providing Buildings/Teacher salary	Types of School	Additional Information
DJ Schools		
AKES / AKES, Community	Kachi, Primary School (5 - 8 years), Kachi, Middle School (8 - 10 years), Kachi, High School (10 + years)	- restricted to Ghizer and Gilgit Districts and to Ismaili areas - aim to increase the enrolment of girls (boys are also admitted) - English now used as medium of teaching - fees moderate - Rs 90-250
Community Schools		
Community / Community	Kachi, Primary School (5 - 8 years), Kachi, Middle School (8 - 10 years), Kachi, High School (10 + years)	- most are English medium - teachers are paid from fees - usually up to primary level but are going to be extended to High School level
Coaching Centres		
Community, AKES, other / NGO's already existing school buildings used	Kachi, Primary School (5 - 8 years), Kachi, Middle School (8 - 10 years), Kachi, High School (10 + years)	- mainly for girls - with financial support might be transformed into "proper" High schools
Private Schools		
Private Investors / Private Investors	Kachi, Primary School (5 - 8 years), Kachi, Middle School (8 - 10 years), Kachi, High School (10 + years)	- often English medium, but also some Urdu medium - founded on charity out of economic, political, religious considerations

provider, accounting for 126 schools and 16% of enrolment, with 35% of all female enrolment. Table 25.2 provides a structural analysis of the major types of schools in the Northern Areas.

The Northern Areas Education Project (NAEP) aims to contribute to the main objective of the World Bank's Northern Education Project (NEP), namely the attainment of a higher enrolment in, and enhanced quality of, elementary education in the Northern Areas, with increased female participation. The specific purpose of NAEP is to achieve sustainable improvements in the management and delivery of elementary education in the NA, especially for girls. The aims will be achieved through:

1) Efficient and effective management by the NA Education Directorate;
2) Improved skills of teacher trainers and teachers;
3) Upgraded textbooks and effective low-cost teaching and learning materials;
4) Effective testing and assessment systems;
5) Strengthened research, monitoring and evaluation capacity;
6) Clear policies and practices for community participation, gender and equity.

The ambitious project addresses a wide range of issues and demands the collaboration and co-operation of all agencies in the Northern Areas. The following factors need to be taken into account.

Ethnic and linguistic diversity

The NA people are an ethnic, linguistic and sectarian kaleidoscope. The main ethnic groups are the Brusho, Mangole and Shin. These peoples have specific geographical locations. Each ethnic group has its own language, the two main languages being Shina (Indic Language) and Balti (Tibetan Language) and with twelve minor languages also being recognized. Urdu is normally used for business or when there is need to communicate between ethnic tribes. Urdu is the mother tongue for only 8% of all the Pakistani people and only 2% of the inhabitants of the NA. The GoP expects all education to be conducted through the medium of Urdu and/or English. This has resulted in all the textbooks being written in Urdu/English, teaching being conducted through these languages. Most of the teachers and pupils do not have a working knowledge of Urdu or English and this causes difficulties in teaching the subject content found in the textbooks. It also follows that if teachers do not have a good command of the teaching language, the pupils' linguistic abilities will be poorly developed. In Government schools the mother tongue replaces Urdu as the teaching language; in English medium schools, Urdu is the teaching language, and not English.

Religion

Education is also influenced by religion and consequently the influence of religious leaders plays an important role. The religion of the NA is Islamic and is heterogeneous with different sects being concentrated in different areas. The three main sects are Shia, Sunni and Ismaili. In some areas the influence of religious leaders proves to be a real obstacle to female education and any changes require the establishment of trustworthy links between NAEP staff and the religious leaders. The problems are greatest in Sunni-dominated districts and least in Ismaili areas.

Social structures

It is sometimes difficult to distinguish between the influence of religion and traditional social structures in education especially in the education of girls. Pakistani society is strongly patriarchal. Women are considered simultaneously to be both of lesser importance than men but also precious. Their role is to be mothers and housekeepers, and are seen to require protection, if necessary with their lives, from other men. This means that women are often prevented from leaving the home unless accompanied by an acceptable male – father, husband or brother. The purdah system may be seen to be either the result of religious belief or as part of the traditional social structure.

Mixed gender schools are quite common at primary level. Difficulties arise at middle or high school levels and girls tend to drop out of school at the age of thirteen. In some parts of NA girls still get married between the ages of 13 - 15 years and parents do not see any reason for investing money in sending their daughters to high school. Where girls have been educated to a higher level, they have some difficulty in finding husbands. The girls themselves wish to marry educated men and also the monetary expectations of the girls' families are much higher – there is a higher bride price. This attitude is changing slowly.

Physical geography

Physical geography influences the decision whether to send children to school. The longest acceptable walking distance for young children at primary level appears to be two to three kilometres. Distance, however, is not the only factor. Where there are obstacles such as cliffs, glaciers or rivers, the acceptable distance is usually much shorter. When boys are older and have outgrown the local education provision it becomes acceptable for them to attend high schools six to eight kilometres distant and sometimes they are sent to stay with relatives or to stay in hostels in Gilgit, Skardu and 'down country'. For girls reaching high school age the situation is more difficult. Often they have to obey strict purdah and are not allowed to leave their villages. If they have no access to middle or high schools at a distance of between two and four kilometres from home and if no transport can be organized, they are often prevented from continuing with higher education.

Parents, especially those who have had the benefit of education, are aware of the need for girls to be educated. Job opportunities, however, in the rural areas are usually limited to the education and the health sectors. This means that if girls are to earn any money they will need to be educated to a certain level to obtain a job in these sectors. Also, boys are becoming more aware of the advantages of having an educated wife. Wives will be able to monitor and help in their children's education if the husband leaves home to seek work.

Economic factors

The awareness of the importance of education is often linked to the economic value a person gives to education. In the NA, girls are an important workforce and have a high economic value for their work on the fields. If they go to school, families have lost part of their work force and are not able to produce as much food. The main function of a daughter is to marry and be a mother. Once a daughter is of marriageable age she will be "lost" to another family and this makes families reluctant to spend too much money on their daughters' education. Most of the farming in NA is subsistence farming. The willingness to spend money on education therefore, also depends on the ability of the family to grow cash crops (fruits, potatoes) and the availability of an income from this source.

Political factors

The influence of politics looms large in the education development of NA. All education activities are subject to the approval of the federal Minister of State in Islamabad and since 1991 the NA Directorate has not been allowed to increase the number of teacher appointments in the areas. The NA

Council, however, does have an important role in determining where schools are opened and who teaches in them. New schools are being opened, but there are no extra teachers available to teach in them. For example, in one valley in Skardu district, 40 schools were established due to political pressure but were not provided with teachers. Political pressure by members of the Council can lead to schools being established in unsuitable locations, teachers being transferred from one school to another resulting in a general reduction of education quality. The transfer of teachers is often politically motivated. Teachers prefer to teach in towns or in their own villages, saving them the expense of transport. In addition, the academic standards of the schools in towns are usually better than in villages. Teachers are often subject to transfer to remote villages, particularly if they have committed some misdemeanour or do not belong to the appropriate political persuasion.

Communications

One of the main difficulties in developing a sustainable education system is the difficulty of communication. The NA is served by daily flights from Islamabad to Gilgit and from Islamabad to Skardu. A complete absence of radar and radio beacons places total reliance on visual navigation and this requires almost perfect flying conditions. Since 1980, the Karakorum Highway has become the NA artery. Landslides and rock falls, floods and mud can interrupt travel on it and can be severed for weeks at a time. Most other roads are un-metalled and passable only by jeep. Distance tends to be calculated in hours rather than kilometres. Landslides and avalanches often block local roads while in winter snow makes roads impassable making educative support highly erratic for many schools. It is also difficult to attract educationalists from other areas of Pakistan to the Northern Areas.

Telephone and facsimile communication is also erratic and dependent on the weather. A rain shower is sufficient to disrupt telecommunications within the NA. Telecommunication within Pakistan is totally unpredictable and organisation of training workshops is difficult.

Terrorist activity

One of the other difficulties in developing an education system in the NA is the threat of terrorist activity. Until recently there had been no activity of this kind in the area but with the bombing of Afghanistan by the USA, terrorist activity has become a reality. This was evident on 13 February, 1999, when the NAEP project office was the target of four bombs.

Centralized directorate of education

The NA has a centralized Directorate of Education controlling all activity in the districts. It has complete power over the distribution of funds, teacher appointments to vacant situations, building of schools and allocation of resources. Each district has its own Deputy Director, but in reality they have very little authority or power. Changes in education are usually implemented from the centre. In Pakistan, the obstacle presented by this centralized power has affected every effort to reform education. When assessing project management, power in NA is most often represented by money. The Directorate is more concerned about the management of funds rather than in the allocation of funds to the districts where it could, if effectively used, provide a positive effect on the standard of education.

Project progress, 1998 – 1999

The NAEP is initially financed for a period of three years with a possible extension of a further two years. Progress to date can be found in Table 25.3. At the present time (1999) the impact of the project has been felt mostly in the Directorate where better organizational systems have been established. Emphasis will in future become district-based. The impact of the project on the quality of pupils' education will be measured using specifically developed achievement tests.

Implications and possible long-term effects of the NAEP

The multiplicity of ethnic groups and religious denominations in this remote region dominate the overall picture making it impossible to treat the area as homogenous. Education in the districts has been influenced both by the GoP and by the attention of international NGO's resulting in education being influenced in many different directions. NAEP is co-ordinating these multifarious efforts and promoting a change in attitudes both at the centre and at the periphery. It involves the acceptance of dynamic education programmes in which education centres are set up, complete their work and replaced by new centres often within short time spans. De-centralization is a central concept for the success of the programme. Directorate staff must be trained to allow a devolution of power from itself to the districts, to the district education officers and ultimately to the schools and other institutions. The schools must be prepared to canvass and accept the ideas and needs of teachers, pupils and parents. All participants must accept the new responsibilities brought about by devolving power to the local areas (Table 25.3).

Table 25.3   Project outputs and progress to date (June, 1999)

Project Outputs	Progress
• Efficient and effective educational management by the Northern Areas Education Directorate;	• Training on team building and decentralized management taken place; GIS used to identify new school sites.
• Improved skills of teacher trainers and teachers;	• Teacher training college sites, lecturers identified; architect plans drawn up.
• Upgraded textbooks and effective low-cost teaching and learning materials;	• Writers identified and trained. Desktop publishing unit established.
• Effective testing and assessment systems;	• Competency lists for all subjects for classes K - 5 developed; Education practitioners trained in item writing methods in Urdu and maths; Baseline tests in Urdu and maths conducted; Continuous Assessment training material developed.
• Strengthened research, monitoring and evaluation capacity;	• Education Management Information System (EMIS) established. Data collection methods developed and improved.
• Clear policies and practices for community participation, gender and equity.	• Policies and practices developed within the above objectives.

The diverse nature of the region demands that no single system or strategy should be developed at the expense of any other. Each local area should be supported to develop sustainable systems and strategies that can be realistically achieved. Attainable targets would enable schools to celebrate success, something, which is rarely achieved at present. All strategies should be focused upon the common goal of providing quality education for all.

The project must accept the overarching role of religion in the society of NA. It is recognized that the one of the reasons for a strengthening of fundamentalism in any religion is due to the feeling of being threatened and vulnerable to the influx of modern ideas into society. Western education development tends to be based more on reasoning and on scientific principles whereas the indigenous religious-based education is based on faith. The project holders must ensure that the development of new education systems and strategies are complementary and not threatening to the religion and culture of the people in the NA. This requires the involvement of the religious leaders in discussions.

Socially, the project will inevitably have wide reaching implications since it requires the greater involvement of girls in education to a higher level. This should result in an better-educated society. Girls will eventually become mothers and they will have higher expectations for their own children's education. This should reduce the rate of illiteracy and innumeracy in the NA.

The improvements in the education system should also increase the economic power of families since the educated children will be able to command better paid occupations. Unless the economic environment of NA changes, this will inevitably lead to outward migration. The more educated people will only be able to find jobs in the cities of Pakistan. This will reduce the workforce available to work on the fields, as once children have been educated the parents feel they would demean themselves by working as peasants.

Eventually, the project could generate some significant political implications. With an increasingly knowledgeable and aware population in the districts, it is probable that there will be an increase in demand for good jobs to be placed in the hands of local people and not to be given to people from outside Northern Areas as is now the case. There may also be a demand for political representation in the Pakistan parliament, which currently does not exist. Improved communication links will be required such as more reliable telephone and e-mail links to allow the region to participate in the life style of the twenty-first century. Economically, the development of an improved quality of education and the development of an educated workforce could result in the area becoming more attractive to entrepreneurs for economic development.

The development of a quality education system in NA is fraught with many difficulties. However, there is a genuine hunger for education improvement amongst the local population. The success of the project will ultimately be judged by the sustained improvement in the quality of the education systems for the local population.

Reference material relevant to the project

British Council (1998), *DFID - Northern Areas Education Project, First Six Monthly Project Progress Report*, 1 May - 31 October, 1998, The British Council, Gilgit.

Merchant, G. and Ali, S. (1999), *Longitudinal Study, Baseline Summary Report*, 1998, NAEP Education Project, British Council and Aga Khan University Institute for Educational Development, The British Council, Islamabad.

PART 5
CONCLUSIONS AND SUMMARY

26 Sustainability as seen from marginal regions – conclusions and summary

HEIKKI JUSSILA, ROSER MAJORAL AND BRADLEY CULLEN

This book focuses on the issues associated with sustainability. It is the outcome of the Annual Meeting of the IGU Commission on Dynamics of Marginal and Critical Regions. The topic of sustainability is an important issue for economic and social development in all regions, especially those that are marginal.

All of the articles included in the book focus on the 'human' aspect of sustainability. This applies not only to those contributions that specifically deal with population or economic development, but also those that analyze the interrelationship between environmental changes and sustainability. A summary of all of the contributions is included in this chapter. Its purpose is to provide a general overview of the book. The editors take full responsibility for the conclusions and any interpretation errors contained in the chapter.

Population, sustainability and marginality

The articles in Part 1 of the book discuss and analyze the impact of population distribution and employment on regional sustainability. They focus on how marginality is connected to sustainability and vice versa. The order of the articles is from the macro (general) level to the micro (local) level.

The first article by Andreoli *et al.* (Chapter 2) describes how the population of Tuscany, Italy became more concentrated between 1861 and 1998. The concentration of population on the lowlands around the River Arno demonstrates how the economic pull of the regions centres gradually 'emptied' the mountainous regions, effectively hampering sustainable economic development. However, the article shows evidence of a turnaround effect during the 1990s, when the most important centres lost population, while some of the accessible mountainous communes gained in population. The article (Chapter 3) by Archer and Lonsdale focuses on the US Midlands. They demonstrate how important communications, in this case the transport system, are to sustaining a region's population. The article shows clearly that the US Interstate Highway system has had a profound effect on the population distribution of the US Midlands. It has enhanced population growth in those counties along its route, while population has stagnated or declined in those communities without direct access to the Interstate. They further argue that:

> *Since amenities are so important to Americans in their locational choices, the most highly favoured places to live and work are the metropolitan areas (more specifically their suburban and ex-urban environs) situated amidst or adjacent to mountains, lakes, seashores, and the like. Those seeking retirement or seasonal recreation have more locational flexibility, but still like to be reasonably close to a metro centre. Thus, areas lacking both amenities and employment diversity are apt to experience ongoing demographic retrogression (Chapter 3, Archer and Lonsdale).*

The focus of Lois Gonzalez's article (Chapter 4) is similar to the article by Archer and Lonsdale. He discusses the depopulation of rural areas in Galicia. But in contrast to the USA, where the growth of large urban centres and indeed migration to these centres from less favourable regions is sometimes seen as positive, depopulation of rural areas in Europe is not always viewed favourably. The basic reasons for the depopulation in Galicia are the same as in Tuscany and the USA – the need to find a job. The migration from Galicia has been to France, Germany and the large urban areas of Spain. The out-migration has been exacerbated by the ecological problems associated with the overuse of the land. The sustainability of these regions of Galicia has, therefore, diminished, which has prompted the development of a variety of programmes within the European Union to promote diversified rural development. The aim is to obtain or regain sustainability in population and economic development.

Regional economic development was also emphasized by Rusanen *et al.* (Chapter 5) in their analysis of the spatial aspects of local poverty in Finland in the mid-1990s. The question of poverty is in many ways fundamental to the notion of sustainability of a region. If one looks at the questions put forward in development theory, it becomes quite evident that without a stable social network that includes equitable incomes, developing a sustainable economic system is difficult. An interesting aspect of this study was that both very densely populated areas and very sparsely populated areas had below average incomes. The 'blessing of population concentration', which is usually a source of economic power, showed its negative side. The 'almost poor' population in most the densely populated areas exceeded the expected figure by a factor of almost three (2.7) (Chapter 5, Rusanen *et al.*). The observations from the Finnish study show a trend towards the spatial concentration of the poor. The income differences were more easily observed when grids rather than regional aggregates, for example, local government districts or NUTS regions, were used. According to the article, poverty should, therefore, be analyzed at as large a scale as possible.

The question of regional imbalances studied by Majoral and Sanchez-Aguilera (Chapter 6) clearly shows that at the regional level sustainability quite often carries an historical 'burden' that is sometimes difficult to overcome. In addition, they demonstrate that policies based on averages have not always been successful, indeed they say that:

The policies implemented in the period of development – the 60s and the beginning of the 70s – were not only unable to reduce regional differences but actually widened the gap. In the last two decades, the policies implemented by the Spanish authorities – at the level of the state and the autonomous communities – and by the EU have aimed, in the main, to reduce the differences in income between the regions (Majoral and Sanchez-Aguilera, Chapter 6).

The result is that the Spanish situation today is characterized by distinct contrasts in income levels, which is a destabilizing element in society. This type of regional structure can also lead to a situation where the general level of sustainability within the society as a whole becomes more fragile. The question of regional sustainability is thus an important issue when moving towards a more integrated and global world.

While Majoral and Sanchez-Aguilera analyze regional imbalances on a national scale, the approach of Muilu (Chapter 7) is local. He analyses two extreme areas in Finland: 1) a growing local area (Seivi), and 2) a declining local area (Suomussalmi). The two areas are very different from each other in the sense that one area's economy (the dynamic and successful case) is based on small private businesses, while the other's (the declining case) relies heavily on public sector jobs. Consequently, a reduced role of state and local governments will have a greater impact on the declining area of Suomussalmi than the growing area of Sievi. The analysis of Muilu shows that, while many remote rural regions are handicapped by poor communications, there are regions that are able to develop a viable, expanding economic base. The positive example of Sievi, as Muilu says, should be an encouragement to other areas, because it is an 'improbable success story'. The issue of sustainability is clear in the case of Sievi, where sustainability is almost a synonym for local will or self-reliance.

The last chapter (Chapter 8) by Pelc analyses rural development and marginal areas in Slovenia. The analysis indicates that if the current depopulation trends continue, it is likely that 'most rural marginal regions (will) become a depopulated wilderness' (Chapter 8, Stanko Pelc). In addition, most areas considered marginal within Slovenia are located on the 'extremes' of the country, i.e., besides being rural and marginal they are also, in relative terms, peripheral to the main centres of Slovenia. The analysis of Pelc shows that while Slovenia is geographically close to 'European Centres,' it is not homogeneous. Consequently, that within the enlarged European Union, Slovenia would not qualify as a single region for Objective 1 (i.e. its GDP is not under 75% of the enlarged EU average). This is a significant impediment to economic development.

But the 'closer to people issues' of landscape, information technologies, recreation, etc. are the issues that can, if used correctly, provide new openings for development. For this reason, it is important to realize that sustainable development does not mean 'blind' importation of methodologies. Methodologies 'have to be adapted to (our) specific conditions and for the specific developmental problems, new authentic methodologies have to be designed.' Therefore, as Pelc points out, it is necessary to look at the people and their needs when aiming at sustainable regional development.

Issues of environment and sustainability

Part II of this book contains six chapters that analyze the environmental aspects of sustainability. Concerns over sustainability are outgrowths of research on environmental issues. It is, therefore, appropriate to devote an entire section of the book to this topic. The interrelationship between sustainability and environmental issues is complex, and has been approached from several angles by the authors in this section of the book.

The first chapter (Chapter 9) by Jones and Al Modayan analyses the problems associated with the conservation of desert environments in Saudi Arabia. They focus on the conflict between the local population and the government. The role of both the people planning for conservation and those living close to these areas have to be taken into account. A sustainable approach is not possible without the co-operation of both locals and planners. The use of a GIS-based methodology can produce an effective tool for land use planning. They argue that:

Modern technology (GIS) has allowed the re-working of much traditional information which has been translated from paper analogue format to digital format. The authors do not recommend GIS as the only means of identifying areas suitable for conservation. Such an approach would be guilty of an oversimplification of a highly variable and complex situation (Jones and Al Modayan, Chapter 9).

Chapter 10 by Leimgruber and Hammer emphasize the importance of people in a sustainable system. The following quote by the UN General Secretary, Kofi Annan, was used to set the stage for their article:

I call on you, individually through your firms, and collectively through your business associations, to embrace support and inact a set of core values in the areas of human rights, living standards and environmental practices (Leimgruber and Hammer, Chapter 10).

Biosphere reserves are approached as sustainable systems in Switzerland. The concept of biosphere is a useful tool when developing an 'active' protection area that does not hinder the economic system. But economic activities must be carried out within the established limits and criteria for the area. The analysis demonstrates biosphere reserves can provide effective models for the development of sustainable systems at a variety of scales. Regional biosphere projects that give 'hope' to a region by providing new tools for local development, also tend to unite the population. The problem at the national level involves developing 'alternative thinking' habits, i.e. an integrative mentality which sees humans and nature as inextricably linked, where sustainable development is not simply a keyword but can be lived (Leimgruber and Hammer, Chapter 10). The idea of biosphere reserves as sustainable systems and, therefore, solutions to the problems of a marginal area will require changes in attitudes. From this point of view, the ideas put forward by Leimgruber and Hammer have much in common with the ideas of articles in the previous section-people do count.

The chapter by Bussing and Norman discuss (Chapter 11) the role of development and environment in Africa. They seek to understand how rural livelihoods can be improved in the drought areas of Botswana. The first important lesson from their analysis is that agriculture holds the key to alleviating poverty in the future. The second lesson is that an agricultural system that does not take into account environmental degradation does not solve the problems of poverty. In this sense, it is imperative that methods and methodologies that stress sustainability be included in the process of development. Development strategies must build upon the strengths and potentialities of the existing systems.

Etienne Nel (Chapter 12) also looks at issues of sustainability from the 'people' or local point of view. The question of self-reliance in an increasingly global environment is analyzed. The challenges that the local self-reliance movement has been facing are many. They range from economic issues to local environmental problems. The basic finding is that local vision, leadership and drive does exist in rural communities. The problem that local communities are facing is that in a global order dominated by free-market and free-trade, there is not much space for those that do not possess special skills and resources to compete. The situation is, according to Nel, quite bleak if the community does not possess a unique resource or skill. In this situation, it is necessary to look for alternative ways for development; ways that can make a difference. In this respect, the role of external agencies becomes important, since they can and are enabling 'bottom-up' initiatives to develop. The approach of Nel supports the idea that a new paradigm of development needs to be developed. A paradigm that is rooted in the culture, ideals and mind-sets of those groups engaged in the development process (Nel, Chapter 12).

Lynch and Pearson analyze (Chapter 13) an issue that is important both from the environmental and economic points of view. The issue of sustainability in Alaska is important, since many people get either additional wealth from hunting, gathering and fishing or they engage in these activities for recreational reasons. In addition, the tourist industry is partly dependent on fish and game. Conflicts over hunting, gathering and fishing are intertwined with broader questions involving how the land should be used. Fishing and hunting is a traditional part of the native population's way of life, while for others the right to hunt and fish is more a question of recreation. In both cases, these interests can conflict with the interests of those trying to develop conservation and preservation plans and others who want to exploit their tourist potential. In other words, as Lynch and Pearson point out, it is possible that the needs of local people to use the animals and other products of the land may come into conflict with the national groups striving for conservation. According to Lynch and Pearson:

> *This issue [conflict] has already arisen about fur trapping which national groups are striving to prevent even though it is a most 'traditional and customary' use of wildlife and for many villagers and some urban dwellers a significant source of cash income (Lynch and Pearson, Chapter 13).*

In an area where land is abundant, the issue is not unused land but land that is subject to restrictions or preferential uses. The question that Lynch and Pearson raise is: Who has the right to decide its usage?

The last chapter (Chapter 14) in this section by Matthews is clearly linked with the previous chapter. It analyses the traditional rights to the land and how they effect modern land use. The question of water rights is a burning issue in an arid region like New Mexico. Besides being vital for life, water is required for economic expansion, as well as the maintenance of traditional lifestyles. The Hispano villagers and Pueblo Indians had developed a system of resource utilization that has in some instances been working for centuries. The United States recognized their water rights. But as the demands of an expanding population for water have increased, conflicts have arisen. Groups who want to ensure economic growth, as well as those who want to protect the natural environment increasingly covet traditional water rights. There is currently a process of litigation that will clarify who has rights to the area's water. When the 'final' verdict is out, there might be a 're-allocation' of water rights, which, according to Matthews, may seriously injure Pueblo communities. A sustainable way of handling natural resources may come to an end.

Global economic development and sustainability

The 'Global economic development and sustainability' section of this book discusses the effects of the global economy on sustainability. The globalizing economic world is an important element that can change the pattern of sustainability. The issue of sustainability from a global and/or economic point of view is more difficult to define, since in overall terms economic sustainability, when measured as economic productivity or profitability, would suggest that large units are more profitable than small units. However, on a regional scale, it is not always so. Small-scale production can and often is the backbone of a region's economy. In fact, it is especially true for most marginal or peripherally located regions. The articles in this section, though the discussions range from consumption, to transport, tourism, fishing, agriculture and indeed city development, all try to answer the same question: How it is possible to obtain sustainable regional development in a globalizing environment?

The first article in this section (Chapter 15) by Mehretu, Pigozzi and Sommers explore marginality and the new international division of labour in more developed countries and less developed countries. The article defines and proposes a typology of marginal regions. They pay particular attention to the exploitation brought about by the new international division of labour. According to the authors, post-Fordist consumption creates a mismatch between the location of production and consumption, which reduces the aggregate demand of the middle class.

The article of Mehretu *et al.* is the most general of the articles in the third section. The articles by Kale (Chapter 16) on the effects of transportation deregulation and Santos-Solla (Chapter 17) on the fishing industry discuss the issue of sustainability within a marginal regional. Kale examines the effect of transportation deregulation on population and unemployment in the US Pacific Northwest (Idaho, Oregon, Washington and western Montana). The time period is from 1970 to 1995, when most of the transportation deregulation took place. He suggests that an analysis of sustain-ability is not possible when using such a short period of time; you need a longer time-span to get a clear 'picture'.

Santos-Solla analyses the economic ramifications of the globalization of the fishing industry. It has resulted in an overuse of fisheries, threatening the entire industry. His study area in Galicia's has been severely impacted, because most of the small coastal towns and villages base their economy on fishing. To stem the tide of over-fishing, international rules and regulations that previously were not present have been imposed. As an Objective 1 region within European Union, this has hurt Galicia. The agreements that are 'good' from the Community's point of view, may, according to Santos-Solla, be 'harmful' for the development of local economies in Galicia. Furthermore, the internationalization of fishing has also benefited large, capital intensive concerns. This has meant that only the bigger fishing ports have been able to develop their activities. In the future, many of the small ports will not be able to compete. In this respect, the analysis by Santos-

Solla shows that a locally sustainable fishing industry may not be compatible with the 'general' aims of the sustainable global fisheries policies.

The two chapters discussing tourism (Chapters 18 and 19), although looking at different sectors, seek to demonstrate the fine balance between economic development and sustainability. The first article by López-Palomeque and Cors-Iglésias looks at 'snow tourism' in Europe. They focus on new approaches to tourism that actively seek to establish a 'sustainable' economic system. The reasoning behind the incorporation of sustainability into tourism is the need to protect natural environments and at the same time to 'secure' economic development and prosperity.

The second tourism based article by Fernandes and Delgado-Cravidão emphasizes rural tourism and its possibilities within the ever-globalizing economy. The questions they ask are: 1. How, within the European context, can rural areas in Portugal find alternative avenues for development? and 2. Is it possible to build a tourist industry on the immaterial and material (manor houses, monasteries, etc.) resources that exist in rural Portugal? The problem is that in Portugal rural tourism may not by itself be enough to revitalize a region's economy. In their words:

It [Rural tourism] can only have a small impact on territory, on pain of disparaging the spirit of innovation / conservation that this activity represents.

Local initiatives sometime need to be connected to 'outside' interests. The chapter (Chapter 20) by Tort-Donada uses the comarca of Priorat as an example of how quality wine production has been used to 'break the evil circle of marginality' in the region. The use of the local label (DO) to market the product has made a difference. The Priorat area has been able to enlarge its economic base, by concentrating on the production of high quality wines. The 'side-effects' have been positive. In the words of Tort-Donanda:

... a marked effect in one particular sense: that of creating, in many villages, a renewed desire to work the land and, above all, to produce wine (Tort-Donada, Chapter 20).

The article (Chapter 21) by Capella-Miternique illustrates how important it is to dream. Changing a destitute, peripheral area is only possible if one has the right kind of imagination and the will to promote dreams. The case of Almeria shows that when a local community unifies around a project many things can happen. In the words of Hugo Capella-Miternique:

> *... it is the strength of our dreams that is most instrumental in development of a region. Dreams place no restrictions on the imagination, they are devoid of prejudices and the most fanciful of ideas can come true (Capella-Miternique, Chapter 21).*

Chapter 22 by Maria José Piñeira describes the problems associated with creating a city that is both attractive and efficient. Economic and population growth that ignores the existing structure and morphology of an urban area can produce environments that are difficult to remedy later. Urban expansion has had its effect on the city of Vigo. An old city originally build in another era needs respect when it grows. The fact that the historical centre of the city is being abandoned is a sign that the previous growth did not embrace the past. The city of Vigo, called the city of 'work' in Galicia, has lost many old buildings. Consequently, the attractiveness of the historical centre has diminished. This is a major problem for Vigo, because, as the author points out, the city centre is where tourists usually congregate in this era of consumption and globalization. It is difficult to see how a city without an interesting centre will be able to survive, even if it is a city of 'work.'

The last chapter (Chapter 23) in the 'Global economic development and sustainability' section of this book by Andreoli and Tellarini analyses several issues facing the Italian farm system. Their work concentrates on the sustainability and economic performance of the system, as well as the use of 'potentially' harmful, polluting substances in agriculture. The analysis is based on economic variables and consequently the results of this analysis must be evaluated in 'monetary' terms. The efficient use of production resources as defined by the authors shows that a large part of the Italian farm system is currently in a state of uncertainty that could lead to abandonment of farming in marginal areas. This could create sustainability problems for local farm communities. In this respect, the EU agricultural and rural policy will have an important effect on sustainability at both the regional and, above all, farm level. The authors conclude by saying that 'policies enabling the reorganization of land use could also allow a more efficient use of farm acreage' and consequently this would lead to a more sustainable and economically more viable farming system in Italy.

Education and sustainable development

The last section of this book, 'Education and sustainable development', contains two chapters, which focus on education's role in developing a sustainable system. Both articles analyze sustainability from the perspective of the individual. Economic growth, change and, above all, sustainability depend on people's ability to use local resources. In this sense, the level of education is an important indicator of a region's 'potential' for developing a genuinely sustainable path for economic and/or social development.

The first chapter (Chapter 24) by Andersson and Blom is more global and yet at the same time regional. By using an approach developed by UNESCO, they show how the 'gap' between the 'educated' developed world and the 'non-educated' developed world increased between 1985-1995. However, although the article initially starts with global issues, its main contribution lies in their discussion of a variety of group's attitudes towards higher education. In global term, all the educational programmes (research studies) discussed in the article have the same aim: to increase the level of education within the particular region. The role of universities and higher education in general is seen as an important tool for development and consequently policies for development tend to integrate these into their programmes. The questions put forward in the article show that in many instances the 'scientific' knowledge arriving from the 'top-down' is not readily appreciated by the local community that prefers 'everyday knowledge' instead of 'scientific knowledge'. This occurs inspite of the fact that, as the case from Sweden shows, local politicians in marginal areas usually are keen on having higher education within the region. Local industries, however, while giving importance to skilled labour, prefer education that they organize rather than 'top-down' higher education. In this respect, the goal of higher education is to both promote development, and to rise the general level of education. The role higher education plays within a region depends, however, on the ties and links the institutions build within that region. Building good links and connections with the people of the region is the best guarantee that the knowledge will effectively diffuse and bring about development that is equitable and sustainable.

The second article (Chapter 25) in this section is by Lucy Jones. Her article discusses the 'implementation of sustainable education' [system] in Pakistan's northern areas, which is among the least-developed areas of Pakistan in terms of education. The educational system is handicapped by the area's culture and physical isolation. One of the main problems in organizing and developing a sustainable education system is that communication is very difficult. The roads are the only 'reliable' means of communication, and they are frequently interrupted by landslides, rock fall, and traffic accidents, making even this type of communication hazardous and sporadic. Under these conditions, according to the author, the educational systems must adapt to the local conditions, but aim at providing quality education for all.

Concluding remarks

This volume on 'Sustainability and marginality in geographical space', provides a glimpse of the IGU Commission on Dynamics of Marginal and Critical Regions' ongoing research on marginality and marginal regions. The focus of this volume, sustainability, has generated significant discussions about the usefulness of the concept and the appropriateness of its use in research not directly related to environmental preservation or conservation. To many, sustainable development is an oxymoron. Since resources are finite, economic development at a global scale is not sustainable. However, as this volume shows, the concept provides a suitable framework for a variety of applications and, when defined properly, it is not only a 'nice word', but also a valuable tool for geographical research.

Printed and bound by CPI Group (UK) Ltd, Croydon, CR0 4YY

22/10/2024

01777605-0006